POLYMER REVIEWS

H. F. Mark *and* **E. H. Immergut,** *Editors*

Volume 1: **The Effects of Ionizing Radiation on Natural and Synthetic High Polymers.** By FRANK A. BOVEY

Volume 2: **Linear and Stereoregular Addition Polymers: Polymerization with Controlled Propagation.** By NORMAN G. GAYLORD and HERMAN F. MARK.

Volume 3: **Polymerization of Aldehydes and Oxides.** By JUNJI FURUKAWA and TAKEO SAEGUSA

Volume 4: **Autohesion and Adhesion of High Polymers.** By S. S. VOYUTSKII

Volume 5: **Polymer Single Crystals.** By PHILIP H. GEIL

Volume 6: **Newer Methods of Polymer Characterization.** Edited by B. KE

Volume 7: **Thermal Degradation of Organic Polymers.** By S. L. MADORSKY

Volume 8: **Metalorganic Polymers.** By. K. A. ANDRIANOV

Volume 9: **Chemistry and Physics of Polycarbonates.** By H. SCHNELL

Volume 10: **Condensation Polymers: By Interfacial and Solution Methods.** By P. W. MORGAN

Volume 11: **Oxidation-Reduction Polymers (Redox Polymers).** By H. G. CASSIDY and K. A. KUN

Volume 12: **Polymerization by Organometallic Compounds.** By L. REICH and A. SCHINDLER

Volume 13: **Polymeric Sulfur and Related Polymers.** By ARTHUR V. TOBOLSKY and WILLIAM J. MACKNIGHT

Volume 14: **Infrared Spectra of Polymers: In the Medium and Long Wavelength Regions.** By DIETER O. HUMMEL

Volume 15: **Stress-Strain Behavior of Elastic Materials: Selected Problem of Large Deformations.** By O. H. VARGA

Volume 16: **Graft Copolymers.** By H. A. J. BATTAERD and G. W. TREGEAR

Volume 17: **High Temperature Resistant Polymers.** By A. H. FRAZER

Volume 18: **Molecular Weight Distributions in Polymers.** By LEIGHTON H. PEEBLES, JR.

Additional volumes in preparation

MOLECULAR WEIGHT DISTRIBUTIONS IN POLYMERS

By LEIGHTON H. PEEBLES, JR.

Chemstrand Research Center, Inc.
a Subsidiary of Monsanto Co.
Durham, North Carolina

INTERSCIENCE PUBLISHERS
A DIVISION OF JOHN WILEY & SONS
NEW YORK • LONDON • SYDNEY • TORONTO

Library of Congress Catalog Card Number: 70-143175

ISBN 0 471 67710 8

Printed in the United States of America

10 9 8 7 6 5 4 3 2 1

INTRODUCTION TO THE SERIES

This series was initiated to permit the review of a field of current interest to polymer chemists and physicists *while* the field was still in a state of development. Each author is encouraged to speculate, to present his own opinions and theories, and, in general, to give his work a more "personal flavor" than is customary in the usual reference book or review article. Whenever background material was required to explain a new development in the light of existing and well-known data, the authors have included them, and, as a result, some of the volumes are lengthier than one would expect of a "review."

We hope that the books in this series will generate as much new research as they attempt to review.

H. F. Mark
E. H. Immergut

PREFACE

In the manufacture of commercial polymers for particular end uses, the question often arises, "How do changes in the process affect the molecular weight distribution of the polymer?" Until recently, this was an almost unanswerable question because the classical methods of determining the distribution were very time consuming. Osmotic pressure measurements for the number-average molecular weight and fractional precipitation for fractionation measurements required extended periods of time to reach equilibrium. The high-speed osmometers and chromatographic fractionation techniques that are now available, however, bring these measurements into the plant laboratory, if not for control purposes, at least for characterization of different polymer lots. With the ready availability of computers, digital and analogue analysis of the distributions and the process variables can easily be used for control or feedback purposes. Because a large number of properties do depend on the molecular weight distribution or on the different averages of the distribution (for an extended list, see Wolf, *Struktur und physikalisches Verhalten der Kunststoffe*, Springer-Verlag, Berlin, 1962), the plant engineer is more able to make intelligent trade-offs on the various properties. The computers can also be used to empirically match mathematical models to actual polymers when the theory or experiment is inadequate.

This book attempts to provide the engineer and the polymer characterization chemist with a summary of the various molecular weight distributions that have been derived from kinetic or statistical arguments in a form that will allow easy comparison of one situation with another. The distribution functions are presented mainly without derivation and with a minimum of commentary on the assumptions used in the derivations, hence the book is rather austere; it is not intended as a definitive text. Individuals interested in the derivations are referred to the original literature. An attempt is made to write the equations using a consistent nomenclature, and to remove errors that occurred in the original papers. Hopefully, more errors are purged than are committed! Most of the graphs of the distribution are drawn for a polymer with a number-average molecular weight of 100. These graphs may be compared with any other degree of polymerization to a good approximation by replacing the abscissa by 100 $r/\bar{r}_n$ and the ordinate by 100 $F(r)/\bar{r}_n$ or 100 $W(r)/\bar{r}_n$, whichever applies. Any error caused by this procedure is probably too small to detect.

I thank Mrs. Faye Collins for typing the original manuscript, Messrs. A. Schlegel, J. Corn, and K. Fussell for aid in the computer calculations, Mrs. Barbara Bassett for preparing the drawings, Mrs. Nancy Poston for her care and patience in typing the final copy and Professor J.J. Hermans for many illuminating discussions.

L. H. PEEBLES, JR.

Raleigh, North Carolina
July, 1970

CONTENTS

Chapter 1

Some General Distribution Functions and Their Properties

Contents

1. Introduction

In the polymerization of vinyl compounds, monomer units are added stepwise to an initiator unit to form long chains of polymer molecules. The initiator units may be either formed instantaneously at the start of polymerization or formed and destroyed during the course of polymerization. Termination and transfer reactions will also alter the number and length of the formed polymer molecules. The total array of molecules is called the distribution. In actuality, it is a discrete function because polymer molecules contain only integer amounts of monomer units.

The frequency function, F(r), is defined as the fraction of molecules of size *r*. It is a normalized function,

$$\sum_{r=1}^{\infty} F(r) = 1 \tag{1}$$

Because the values of *r* used in (1) are significantly larger than unity, the discrete function may be replaced by a continuous function F(r) dr, and it too is normalized.

$$\int_0^{\infty} F(r)\, dr = 1 \tag{2}$$

The integral may go from zero instead of unity because F(0) is either extremely small or zero. In the continuous function, it is understood that F(r) dr is that fraction of molecules lying between *r* and r + dr. Because *dr* is unity for molecules, we tend to think of F(r) (where *r* is specified) as always being a fractional number. However, there are instances when F(y) is considerably larger than unity and *dy* a very small number, such as when compositional distributions are considered.

On occasions the concentration of molecules of size *r*, P_r, will be given whenever the normalized form, F(r), is unknown, that is, the total concentration of polymers is unknown.

The weight fraction of molecules of size *r* is given by

$$W(r) = rF(r) / \sum_{r=1}^{\infty} rF(r) \tag{3}$$

$$W(r)\,dr = rF(r)\,dr / \int_{-\infty}^{+\infty} rF(r)\,dr \tag{4}$$

The nth moment of a distribution is given by

$$\mu_n = \int_{-\infty}^{+\infty} r^n F(r)\,dr \tag{5}$$

$$= \sum_r r^n F(r) \tag{6}$$

In molecular weight distributions negative values of *r* have no meaning so the lower limit of the integral sign can be replaced by zero, but for the moment we retain the infinite negative limit to ensure complete generality. The sum over *r* means over all permissible values of *r*.

The *mean* of a distribution is

$$\mu_1 = \int_{-\infty}^{+\infty} rF(r)\,dr \tag{7}$$

$$= \sum_r rF(r) \tag{8}$$

while the <u>variance</u> of a distribution is

$$\sigma^2 = \mu_2 - \mu_1^2 = \int_{-\infty}^{+\infty} (r - \mu_1)^2 F(r)\, dr \tag{9}$$

$$= \sum_r (r - \mu_1)^2 F(r) \tag{10}$$

The <u>average</u> of a distribution is defined by

$$\bar{r}_i = \sum_r r^i\, F(r) / \sum_r r^{i-1} F(r) \tag{11}$$

$$= \int_{-\infty}^{+\infty} r^i F(r)\, dr \Big/ \int_{-\infty}^{+\infty} r^{i-1} F(r)\, dr \tag{12}$$

$$= \mu_i / \mu_{i-1} \tag{13}$$

The number-average degree of polymerization is $\bar{r}_1$, but we denote it by $\bar{r}_n$,

$$\bar{r}_n = \sum_{r=1}^{\infty} rF(r) / \sum_{r=1}^{\infty} F(r) \tag{14}$$

$$= \sum_{r=1}^{\infty} W(r) / \sum_{r=1}^{\infty} [W(r)/r] \tag{15}$$

The weight-average degree of polymerization is $\bar{r}_2$, which we denote by $\bar{r}_w$,

$$\bar{r}_w = \sum_{r=1}^{\infty} r^2 F(r) / \sum_{r=1}^{\infty} rF(r) \tag{16}$$

$$= \sum_{r=1}^{\infty} rW(r) / \sum_{r=1}^{\infty} W(r) \tag{17}$$

The weight average is always larger than the number average because the larger species are counted more heavily (r^2 vs. $\underline{r}$).

These averages can be substituted into equation (9) to give

$$\sigma^2 = \bar{r}_w \bar{r}_n - \bar{r}_n^{\,2} \tag{18}$$

$$= \bar{r}_n^{\,2} [(\bar{r}_w / \bar{r}_n) - 1] \tag{19}$$

$\bar{r}_w > \bar{r}_n \therefore \left(\frac{\bar{r}_w}{\bar{r}_n} - 1 \right) > 0.$

This measure of the breadth of a distribution will always increase as $\bar{r}_n$ increases except for a unimolecular distribution, $\bar{r}_w / \bar{r}_n = 1$. For this reason, we prefer to measure the breadth of a molecular weight distribution by the dispersion ratio, $\bar{r}_w / \bar{r}_n$.
(uneinheitlichkeit, nonuniformity coefficient)

From the definitions of F(r) and W(r) and $\bar{r}_n$

$$W(r) = rF(r)/\bar{r}_n \tag{20}$$

The number-average degree of polymerization is the number-average molecular weight divided by the molecular weight of the repeat unit. Rewriting equation (14) in terms of polymer concentration, we have

$$\bar{r}_n = \sum_{r=1}^{\infty} rP_r / \sum_{r=1}^{\infty} P_r$$

We see immediately that the number-average molecular weight of a polymer is merely the ratio of total weight of polymer to the total number of polymer molecules. In many instances, it is theoretically easier to calculate the total weights and numbers than to perform the required summations.

The *z* and z + 1 averages are defined by equations (11) and (12) with *i* equal to 3 and 4, respectively. There is no need to restrict the averages of a distribution to positive integers--any useful average can be defined; such as the (-5/2) average. The intrinsic viscosity of a polymer is related to its molecular weight through the equation

$$[\eta] = K\bar{r}_v^{\,a} \tag{21}$$

where $\bar{r}_v$ is the viscosity-average degree of polymerization. It is related to the frequency function by

$$\bar{r}_v = \left[\sum_{r=1}^{\infty} r^{1+a}F(r) / \sum_{r=1}^{\infty} rF(r) \right]^{1/a} \tag{22}$$

and to the weight function by

$$\bar{r}_v = \left[\sum_{r=1}^{\infty} r^a W(r) / \sum_{r=1}^{\infty} W(r) \right]^{1/a} \tag{23}$$

In principle, a distribution function can be determined if sufficient averages of the distribution can be determined. In practice, only the number, weight, viscosity, and perhaps the _z_ averages can be found, which are insufficient to define any distribution without making further assumptions. In the sections that follow we examine some one-, two-, and three-parameter distribution equations which find use either because of their general applicability to polymeric systems or the ease of determining the parameters from measurements on polymer fractions.

2. The Schulz-Flory Most Probable Distribution.--A one-parameter equation

When a linear addition polymer is formed by a constant rate of initiation, monomer concentration invariant, transfer to solvent but not to monomer, and termination by disproportionation; or when a linear condensation polymer is formed by assuming equal reactivity of all chain ends; or when a linear condensation polymer is formed by allowing the units to interchange in a random manner; or when a low molecular weight linear polymer is formed from a higher molecular weight linear polymer by random scission--the resulting molecular weight distribution is

$$F(r) = p^{r-1}(1 - p) \tag{1}$$

$$W(r) = rp^{r-1}(1 - p)^2 \tag{2}$$

(Schulz, 1; Flory, 2, 3) where, in condensation polymerization, p is the extent of reaction. Because so many systems appear to obey this distribution, it is called the "most probable distribution." We use this function as a standard for comparison with other, more complex distributions. The number-, weight-, and z-average degrees of polymerization are

$$\bar{r}_n = 1/(1 - p) \tag{3}$$

$$\bar{r}_w = (1 + p)/(1 - p) \tag{4}$$

$$\bar{r}_z = (1 + 4p + p^2)/(1 - p)(1 + p) \tag{5}$$

and the viscosity-average degree of polymerization is

$$\bar{r}_v = \bar{r}_n[\Gamma(2 + a)]^{1/a} \tag{6}$$

where a is given by equation (1.21). Note that the ratios of the degrees of polymerization for high molecular weight polymer are

$$\bar{r}_n:\bar{r}_w:\bar{r}_z = 1:2:3$$

The integral distribution curve is

$$\int_0^r W(r)\ dr = \frac{(1 - p)^2}{(\ln p)^2} \{(1/p) + [(r \ln p) - 1]p^{r-1}\} \quad (7)$$

$$\approx (1/p) - [1 + (1 - p)r]p^{r-1} \quad (8)$$

with the approximation $\ln p \approx p - 1$. The integral frequency distribution can be linearized

$$\log [1 - \int_0^r F(r)\ dr] = (r - 1) \log p \quad (9)$$

(Fairnerman and Polykova , 4). Curves of F(r) and W(r) as a function of <u>r</u> are given in Figures 1.2.1 and 1.2.2 for various values of $\bar{r}_n$.

Although the curves appear to be significantly different from one another, they differ in fact only by a scaling factor. Thus equation (2) can be written as

$$W(r) = pr(1 - p)^2 p^r \quad (10)$$

$$= pr(1 - p)^2 \exp(ar) \quad (11)$$

where $a = \ln p$

With equation (3)

$$\bar{r}_n = 1/(1 - p) \approx -1/(\ln p) \quad (12)$$

which is valid provided that $(1 - p)^2/2 \ll (1 - p)$ we can write

Curve	a	b	c	d
$\bar{r}_n$	50	100	200	400
$\bar{r}_w$	99	199	399	799

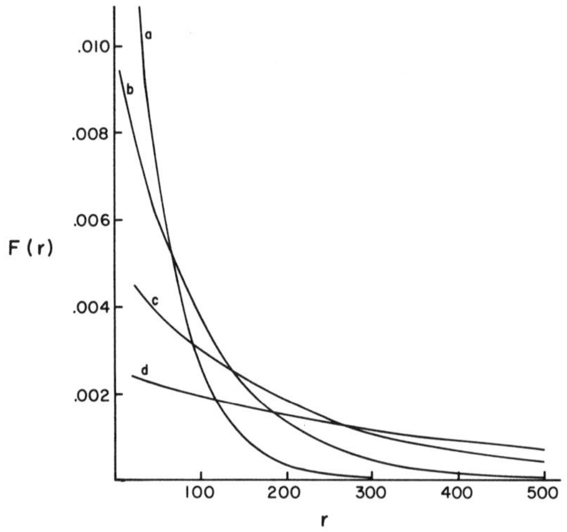

Figure 1.2.1 - The Schulz-Flory "most probable" distribution, equation 1.2.1 as a function of r for various values of $\bar{r}_n$. (Courtesy J. Am. Chem. Soc.) (Flory).

$$W(r)\cdot\bar{r}_n \approx p(r/\bar{r}_n)\ \exp(-r/\bar{r}_n) \qquad (13)$$

$$\approx (r/\bar{r}_n)\ \exp(-r/\bar{r}_n) \qquad (14)$$

since $p \approx 1$.

Similarly

$$F(r)\cdot\bar{r}_n \approx \exp(-r/\bar{r}_n) \qquad (15)$$

Curve	a	b	c	d
$\bar{r}_n$	50	100	200	400

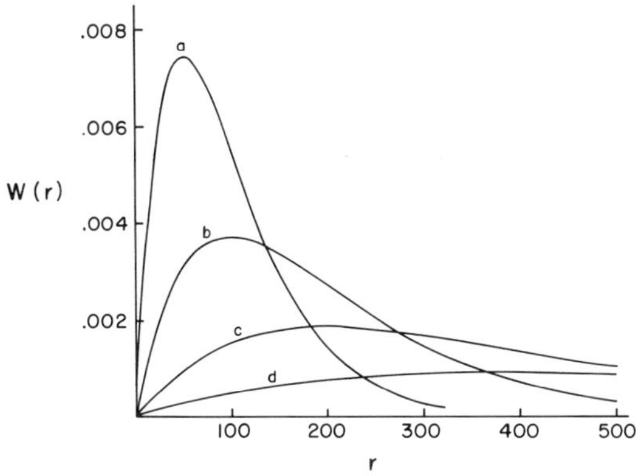

Figure 1.2.2 - The Schulz-Flory "most probable" distribution, equation 1.2.2 as a function of *r* for various values of $\bar{r}_n$. (Courtesy *J. Am. Chem. Soc.*) (Flory, 2).

hence the alternate names of "geometric distribution function" or "exponential distribution function."

The maximum in the weight fraction distribution occurs at $\bar{r}_n$, while the inflection point is at $\bar{r}_w$.

3. The Schulz Distribution.--A two-parameter equation

In 1935 Schulz (1) derived the following equation for polymerization with a constant rate of initiation and termination by second-order interaction with monomer

$$W(r) = (-\ln p)^2 r p^r \tag{1}$$

(Schulz, 1) which is equivalent to the Schulz-Flory function, equation (2.2), provided that $(-\ln p) \approx 1 - p$.

In 1939 equation (1) was generalized to

$$F(r) = (-\ln p)^k r^{k-1} p^r / \Gamma(k) \tag{2}$$

$$W(r) = (-\ln p)^{k+1} r^k p^r / \Gamma(k + 1) \tag{3}$$

(Schulz, 5; Zimm, 6). When $k = 2$, the distribution function is that derived for addition polymerization with a constant rate of initiation, monomer concentration invariant, no transfer, and termination by coupling of active molecules. Figures 1.3.1 and 1.3.2 present curves of $F(r)$ and $W(r)$ as a function of $\underline{r}$ and $\bar{r}_n$ for $k = 2$. Note that in contrast to the Schulz-Flory distribution, $F(r)$ has a maximum at $\bar{r}_n/2$. Curves for $F(r)$ and $W(r)$ for $\bar{r}_n = 100$ are presented in Figures 1.3.3 through 1.3.6 as a function of $\underline{r}$ and $\underline{k}$. As $\underline{k}$ increases, the distribution becomes narrower. The maximum in the $F(r)$ distribution occurs at $(k - 1)\bar{r}_n/k$, while the maximum in $W(r)$ always occurs at $\bar{r}_n$.

The various average degrees of polymerization are

$$\bar{r}_n = -k/\ln p \tag{4}$$

$$\bar{r}_w = -(k + 1)/\ln p \tag{5}$$

$$\bar{r}_z = -(k + 2)/\ln p \tag{6}$$

Curve	a	b	c	d
$\bar{r}_n$	50	100	200	400
$\bar{r}_w$	75	150	300	600

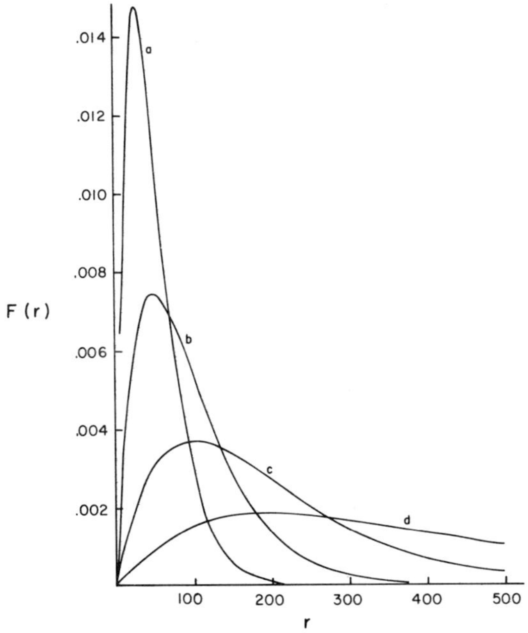

Figure 1.3.1 - The frequency function for the Schulz distribution with k = 2, as a function of r for various values of $\bar{r}_n$. (after Schulz, 5)

$$\bar{r}_v = \frac{\bar{r}_n}{k}\left[\frac{\Gamma(1 + k + a)}{\Gamma(1 + k)}\right]^{1/a} \tag{7}$$

where a is given by equation (1.21). The ratios of the degrees of polymerization for high molecular weight polymer are

Curve	a	b	c	d
$\bar{r}_n$	50	100	200	400

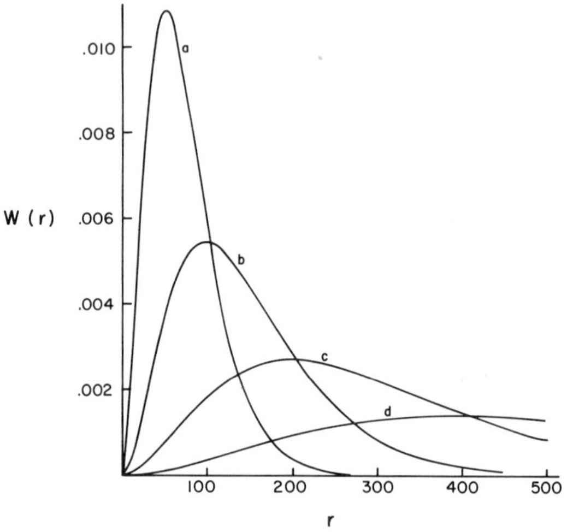

Figure 1.3.2 - The weight function for the Schulz distribution. Same conditions as in Figure 1.3.1. (after Schulz, 5).

$$\bar{r}_n : \bar{r}_w : \bar{r}_z = k:(k+1):(k+2)$$

The cumulative number or weight fraction may be computed from

$$\int_0^r F(r)\,dr = \frac{k(-\ln p)^k}{\Gamma(1+k)} \sum_{i=0}^{\infty} \frac{(-1)^i(-\ln p)^i r^{k+i}}{i!\,(k+i)} \tag{8}$$

$$\int_0^r W(r)\,dr = \frac{(-\ln p)^{k+1}}{\Gamma(1+k)} \sum_{i=0}^{\infty} \frac{(-1)^i(-\ln p)^i r^{k+i+1}}{i!\,(k+i+1)} \tag{9}$$

k	1	2	4	6	10
$\bar{r}_w$	200	150	125	116.7	110

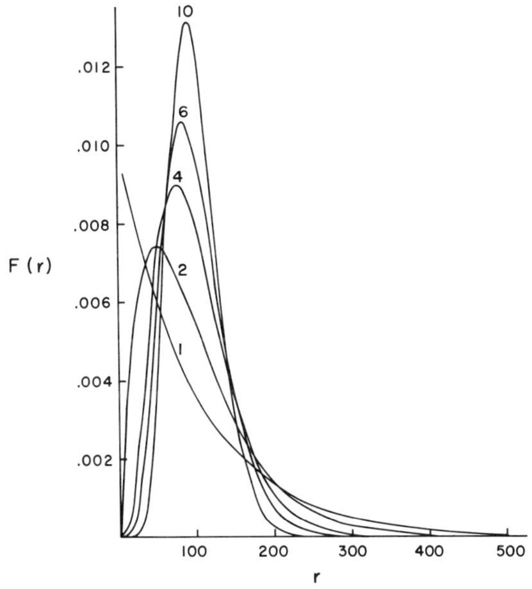

Figure 1.3.3 - The frequency function of the Schulz distribution with $\bar{r}_n = 100$ as a function of $\underline{r}$ for $k \geqslant 1$. The distribution becomes narrower with increasing values of $\underline{k}$. (after Schulz, 5)

Kotliar (7) notes that although the Schulz distribution may provide an adequate fit to fractionation data, the determination of $\bar{r}_n$ and $\bar{r}_w$ from the distribution may be greatly in error.

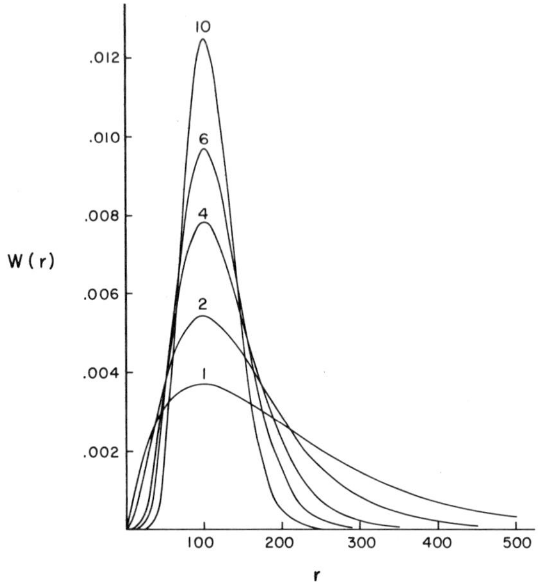

Figure 1.3.4 - The weight function of the Schulz distribution. Same conditions as in Figure 1.3.3. (after Schulz, 5)

4. The Generalized Exponential Distribution.--A three-parameter equation

The preceding distributions are special cases of the generalized exponential distribution:

$$F_{m,k,y}(r) = my^{k/m}\, r^{k-1}[\exp(-yr^{m})]/\Gamma(k/m) \tag{1}$$

$$W_{m,k,y}(r) = my^{(k+1)/m}\, r^{k}[\exp(-yr^{m})]/\Gamma[(k+1)/m] \tag{2}$$

k	2	1	1/4	1/9	1/24
$\bar{r}_w$	150	200	500	1000	2500

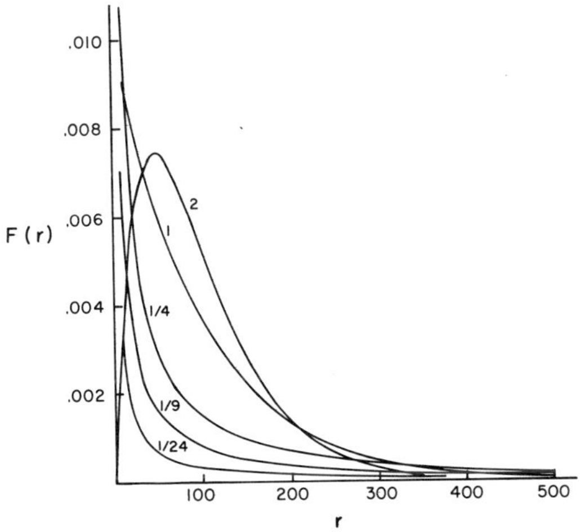

Figure 1.3.5 - The frequency function of the Schulz distribution with $\bar{r}_n = 100$ as a function of r for $k \leqslant 2$. The distribution becomes wider with decreasing values of k. (after Schulz, 5)

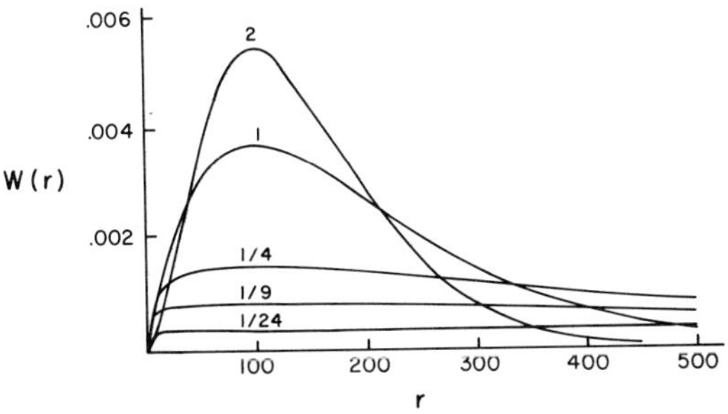

Figure 1.3.6 - The weight function of the Schulz distribution. Same conditions as in Figure 1.3.5. (after Schulz, 5)

(Muus and Stockmayer, 8) where $y = -\ln p$ for the Schulz distribution. The average degrees of polymerization are

$$\bar{r}_n = \Gamma[(k + 1)/m]/y^{1/m} \; \Gamma[k/m] \tag{3}$$

$$\bar{r}_w = \Gamma[(k + 2)/m]/y^{1/m} \; \Gamma[(k + 1)/m] \tag{4}$$

$$\bar{r}_z = \Gamma[(k + 3)/m]/y^{1/m} \; \Gamma[(k + 2)/m] \tag{5}$$

$$\bar{r}_i = \Gamma[(k + i)/m]/y^{1/m} \; \Gamma[(k + i - 1)/m] \tag{6}$$

and

$$\bar{r}_v = \{\Gamma[(k + a + 1)/m]/y^{a/m} \; \Gamma[(k + 1)/m]\}^{1/a} \tag{7}$$

where a is defined by equation (1.21).

The Weibull or Tung distribution is obtained by setting $k = m - 1$ and $k > 0$. This distribution is usually seen in the form

$$\int_0^r W(r)\, dr = I(r) = 1 - \exp\,[-yr^m] \tag{8}$$

(Weibull, 9; Tung, 10) where m and y are found by plotting the logarithm of r against log log $\{1/[1 - I(r)]\}$. See Figure 1.4.3. The Weibull-Tung distribution is shown in Figures 1.4.1 and 1.4.2 for $\bar{r}_n = 100$ and various values of $\bar{r}_w$.

Curve	a	b	c	d
$\bar{r}_w$	110	150	200	500

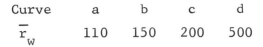

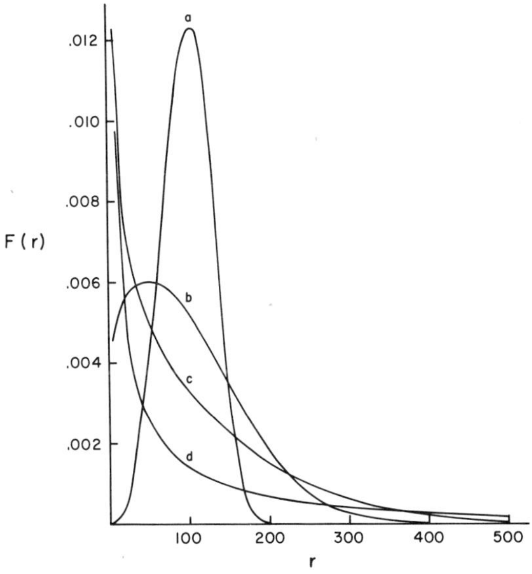

Figure 1.4.1 - The frequency function for the Weibull-Tung distribution with $\bar{r}_n = 100$ as a function of *r*.

5. The Normal Distribution.--A two-parameter equation

The normal distribution function or Gaussian distribution is the bell-shaped distribution for individuals symmetrically distributed about the mean:

$$F(r) = \frac{\exp\{-(r - \bar{r})^2/2\sigma^2\}}{(2\pi)^{\frac{1}{2}}\sigma} \qquad (1)$$

Curve	a	b	c	d
$\bar{r}_w$	110	150	200	500

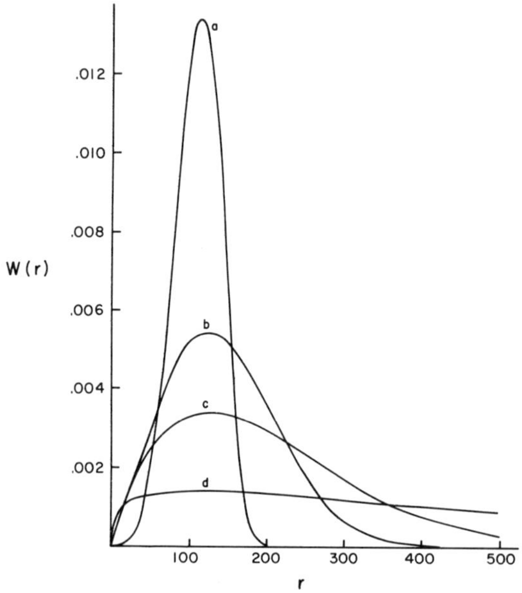

Figure 1.4.2 - The weight function for the Weibull-Tung distribution with $\bar{r}_n$ = 100 as a function of r.

where σ is the half width of the distribution measured at one-half the maximum height of the distribution. If the number of molecules is normally distributed, then the weight distribution is given by

$$W(r) = (r/\bar{r})F(r) \tag{2}$$

The mean of the distribution has the same meaning as the number-average degree of polymerization

$$I(r) = \int_0^r W(r)\, dr$$

$\bar{r}_n$	$\bar{r}_w$	m	y
100	110	4.25	1.427×10^{-9}
100	147	2.15	1.674×10^{-5}
100	202	1.65	1.305×10^{-4}
100	499	1.21	5.030×10^{-4}

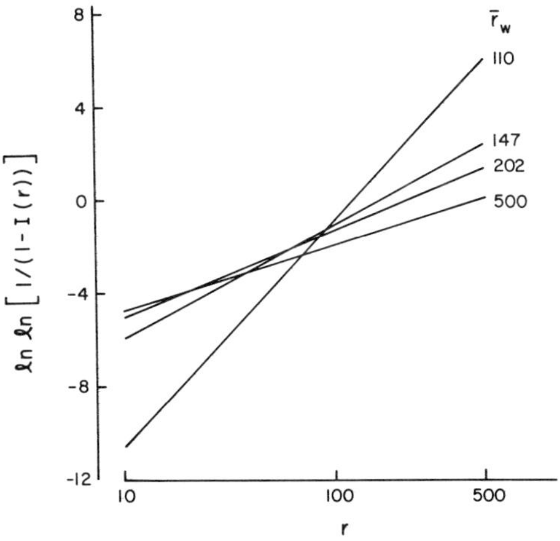

Figure 1.4.3 - Linearization of the Weibull-Tung distribution.

$$\bar{r} = \int_{-\infty}^{+\infty} rF(r)\, dr = \bar{r}_n \tag{3}$$

but the integration must cover negative values of r. The variance of the distribution is

$$\sigma^2 = \int_{-\infty}^{+\infty} (r - \bar{r})^2 F(r)\, dr \tag{4}$$

and the weight-average degree of polymerization is

$$\bar{r}_w = \frac{\sigma^2}{\bar{r}_n} + \bar{r}_n \tag{5}$$

F(r) and W(r) are plotted in Figure 1.5.1, curves a and b for $\bar{r}_n = 100$, $\bar{r}_w/\bar{r}_n = 2$ for the case of a normally distributed number fraction. A significant fraction of the area under the F(r) curve extends to the left of zero, while the area under W(r), curve b, is greater than unity.

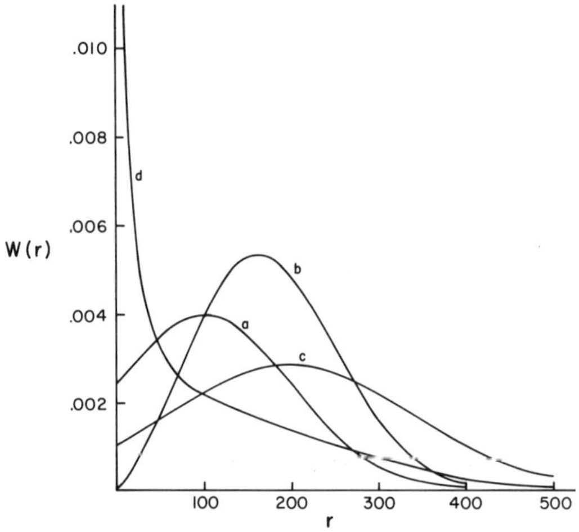

Figure 1.5.1 - Comparison of the frequency and weight functions for the normal distribution when $\bar{r}_n = 100$ and $\bar{r}_w = 200$. a F(r) is distributed normally about 100 and the corresponding weight curve b. c W(r) is distributed normally about 200 and the corresponding frequency curve d. To obtain unit areas, the integration limits are $-\infty$ to $+\infty$ rather than the usual 0 to $+\infty$.

If, on the other hand, we take the normal curve to be a representation of the weight-fraction distribution, then $\bar{r}$ in equation (1) would be $\bar{r}_w$. A plot of this type of distribution is given by curve c with $\bar{r}_n = 100$, $\bar{r}_w = 200$, and again a significant fraction of the area under the curve extends to the left of zero. The frequency function curve F*(r) is thus

$$F^*(r) = \bar{r}_n F(r)/r \tag{6}$$

which is shown by curve d. For polymers, the normal curve can be used only for very narrowly distributed systems as an approximation of the Poisson distribution or polymers with a somewhat wider distribution.

6. The Logarithmic Normal Distribution.--A two-parameter equation

The normal distribution function, given above, cannot be used to describe a broad molecular weight distribution because negative molecular weights do not exist. To avoid this problem and yet be able to use the known properties of the normal distribution, we can assume that the weight distribution of the logarithm of molecular size is normally distributed. Thus we replace r in equation (5.1) by ln r and $\bar{r}$ by ln $\bar{r}_m$

$$W(\ln r) = \frac{\exp\{-(\ln r - \ln \bar{r}_m)^2/2\sigma^2\}}{(2\pi)^{\frac{1}{2}}\sigma} \tag{1}$$

$$\int_1^\infty W(\ln r)\, d(\ln r) = 1 \tag{2}$$

or in the alternate form

$$W(r) = \frac{\exp\{-(\ln r - \ln \bar{r}_m)^2/2\sigma^2\}}{r(2\pi)^{\frac{1}{2}}\sigma} \tag{3}$$

$$\int_0^\infty W(r)\, dr = 1 \tag{4}$$

(Wesslau, 11) where

$$\ln \bar{r}_m = \int_0^\infty (\ln r)\, W(r)\, dr \tag{5}$$

Here $\bar{r}_m$ is the median value of the distribution, that is, one half of the values of r are less than $\bar{r}_m$. The parameter σ is the standard deviation of ln r.

$$\sigma^2 = \int_0^\infty (\ln r - \ln r_m)^2 W(r)\, dr \tag{6}$$

The average degrees of polymerization are

$$\bar{r}_n = \bar{r}_m \exp(-\sigma^2/2) \tag{7}$$

$$\bar{r}_w = \bar{r}_m \exp(+\sigma^2/2) \tag{8}$$

$$\bar{r}_z = \bar{r}_m \exp(+3\sigma^2/2) \tag{9}$$

$$\bar{r}_i = \bar{r}_m \exp\{(2i - 3)\sigma^2/2\} \tag{10}$$

from which we can show that

$$\bar{r}_m = (\bar{r}_n \bar{r}_w)^{\frac{1}{2}} \tag{11}$$

$$\bar{r}_w/\bar{r}_n = \bar{r}_z/\bar{r}_w = \bar{r}_{z+1}/\bar{r}_z = \exp \sigma^2 \tag{12}$$

The maximum of the weight distribution function, W(r), is located at $\bar{r}_n^{3/2}/\bar{r}_w^{1/2}$ while the maximum of the frequency distribution, F(r), is located at $\bar{r}_n^{5/2}/\bar{r}_w^{3/2}$. Equation (3) has been given in a generalized, three-parameter form:

$$W_s(r) = \frac{r^s \exp[-(\ln r/\bar{r}_s)^2/2\sigma^2]}{(2\pi)^{\frac{1}{2}}\sigma\bar{r}_s^{s+1}\exp\{(s+1)^2\sigma^2/2\}} \tag{13}$$

(Espenshied, Kerker, and Matejewic, 12) where $\bar{r}_s$ is a constant, related to the various molecular weight averages by

$$\bar{r}_n = \bar{r}_s \exp\{(2s+1)\sigma^2/2\} \tag{14}$$

$$\bar{r}_w = \bar{r}_s \exp\{(2s+3)\sigma^2/2\} \tag{15}$$

$$\bar{r}_z = \bar{r}_s \exp\{(2s+5)\sigma^2/2\} \tag{16}$$

$$\bar{r}_i = \bar{r}_s \exp\{[2(s+i)-1]\sigma^2/2\} \tag{17}$$

These equations are also related by equation (12). Honig (13) has shown that equation (13) can be transformed back into equation (3) by means of the relation

$$\ln \bar{r}_m = \ln \bar{r}_s + (s + 1)\sigma^2 \tag{18}$$

Indeed, calculation of various distribution curves from equation (13) with assorted values of <u>s</u> lead to exactly the same curve. Thus the normalized Lansing-Kramer function (s = 0) (Lansing and Kramer, 14) is identical to the normalized Wesslau function (s = -1) (Wesslau, 11). The frequency and weight distributions for $\bar{r}_n = 100$ and various values of $\bar{r}_w/\bar{r}_n$ are shown in Figures 1.6.1 and 1.6.2. The show the symmetrical shape, W(r) is plotted against log r in Figure 1.6.3. Kotliar has shown that the logarithmic normal distribution function is not a good representation of a polymer after either low or high molecular weight material is removed or degradation has occurred (15, 16) nor is it a good method of evaluating $\bar{r}_w/\bar{r}_n$ (7).

To find $\bar{r}_m$ and σ of the logarithmic normal distribution for a fractionated polymer, plot the cumulative weight fraction of polymer as the ordinate on "probability paper," and ln r as the abscissa. Draw the best straight line through the points. At P = 0.50, $\ln r = \ln \bar{r}_m$; at $P = 0.500 \pm 0.341$, $\ln r - \ln \bar{r}_m = \pm \sigma$. See Figure 1.6.4.

The Weibull-Tung distribution and the logarithmic normal distribution both find use because of the ease in determining their parameters. Figures 1.6.5 and 1.6.6 compare the frequency and weight distributions of these functions and the Schulz distribution for $\bar{r}_n = 100$ and various values of $\bar{r}_w$.

Curve	a	b	c	d
$\bar{r}_w$	110	150	200	500

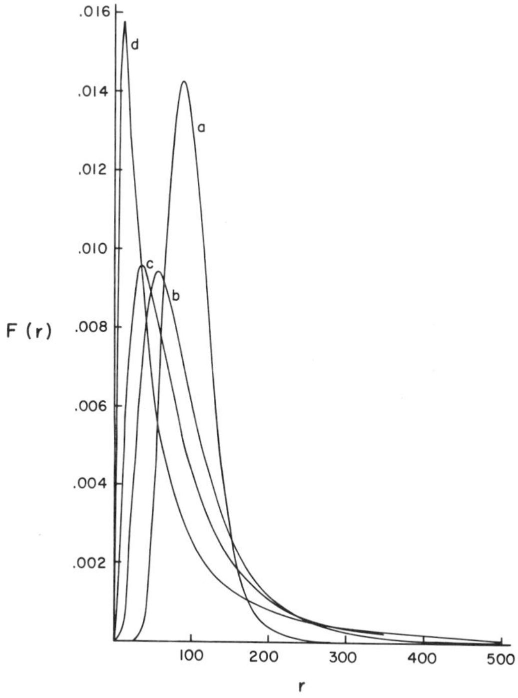

Figure 1.6.1 - Frequency function for the logarithmic normal distribution function as a function of r for $\bar{r}_n$ = 100.

7. The Poisson Distribution.--A one-parameter equation

If a number of items, n, are to be distributed randomly into a number of categories c, n > c, the expected number of items per category is $\nu = n/c$ and the frequency function for r items per category is

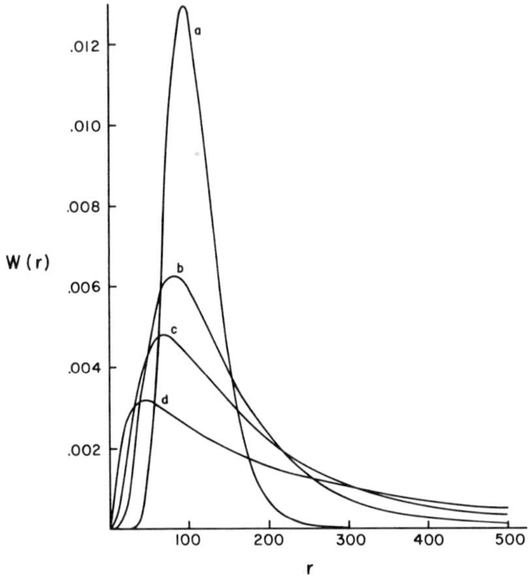

Figure 1.6.2 - Weight fraction distribution for the logarithmic normal distribution function. Same conditions as in Figure 1.6.1.

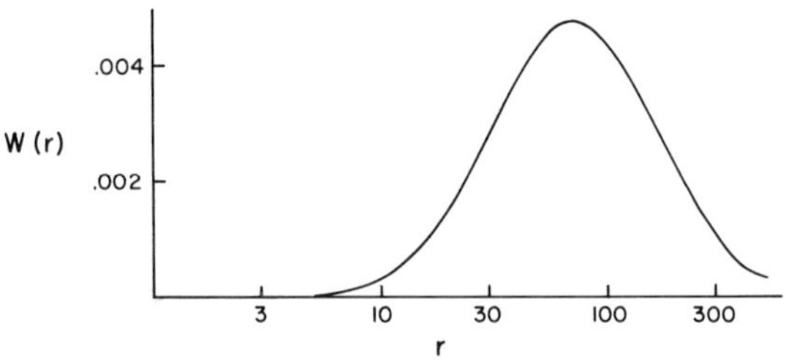

Figure 1.6.3 - Logarithmic normal distribution as a function of log r with $\bar{r}_n = 100$, $\bar{r}_w/\bar{r}_n = 2$ to show the bell-shaped distribution.

$\bar{r}_n$	$\bar{r}_w$	$\bar{r}_m$	σ
100	110	104.9	.309
100	150	122.5	.637
100	200	141.5	.832
100	500	223.6	1.268

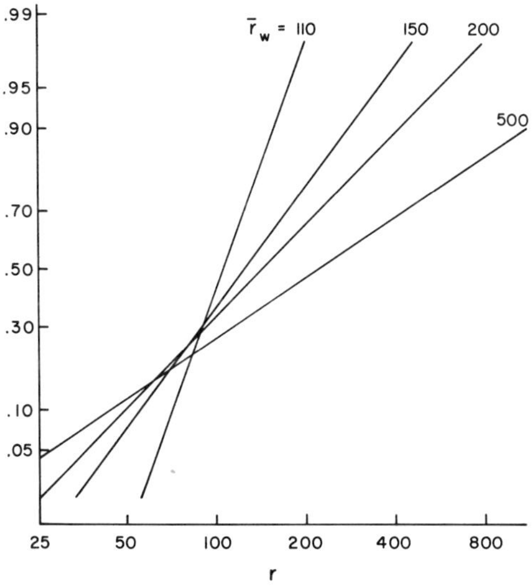

Figure 1.6.4 - Linearization of the logarithmic normal distribution on "probability paper."

$$F(r) = [\exp(-\nu)]\nu^r/r! \qquad (1)$$

where

$$n = \sum_c r \qquad (2)$$

If there are c initiator fragments and r monomer units and if the initiator fragments are counted as part of the polymer chain, the minimum value of r must be unity, not zero; hence

$$F(r) = [\exp(-\nu)]\nu^{r-1}/(r - 1)! \quad (3)$$

$$\bar{r}_n = 1 + \nu \quad (4)$$

$$\bar{r}_w = (1 + 3\nu + \nu^2)/(1 + \nu) \quad (5)$$

$$\approx 1 + \bar{r}_n$$

For all practical purposes, there is essentially no difference between F(r) and W(r) (see Figure 3.1.1, page 137). Figure 1.7.1 compares the Poisson and Schulz-Flory distributions for $\bar{r}_n = 100$, while Figure 1.7.2 gives W(r) for various values of ν. Again, although the breadth of the distribution, measured as σ, does increase with increasing ν, the dispersion ratio decreases with increasing ν.

8. Bimodal Distributions

It is instructive to examine what kind of distributions are bimodal. Let us consider polymers which conform to the Schulz-type of distribution:

$$W(r) = (-\ln p)^{k+1} r^k p^r/\Gamma(k + 1) \quad (1)$$

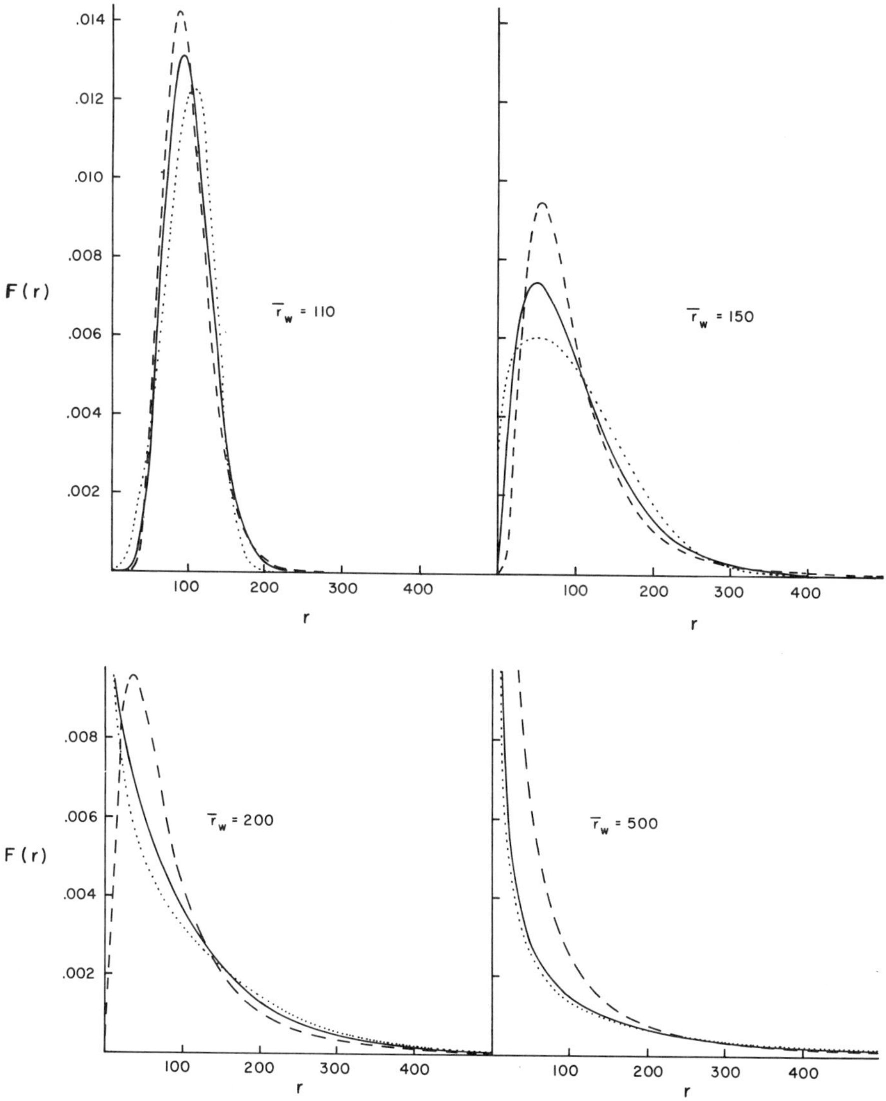

Figure 1.6.5 - Comparison of the frequency functions of the Schulz (———), Weibull-Tung (— — —), and logarithmic normal (······) distributions as a function of $\underline{r}$ for $\overline{r}_n = 100$ and the indicated values of $\overline{r}_w$.

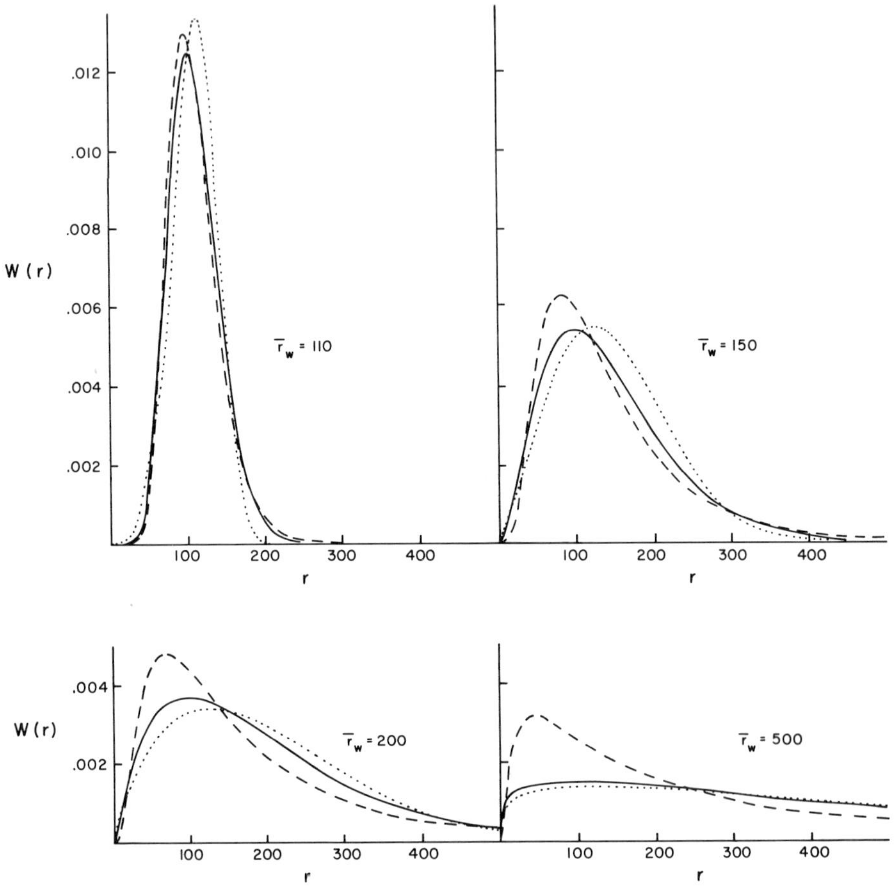

Figure 1.6.6 - Comparison of the weight functions of the Schulz (———), Weibull-Tung (— — —), and logarithmic normal (······) distributions as a function of r for $\bar{r}_n$ = 100 and the indicated values of $\bar{r}_w$.

or, in its alternate form,

$$W(r) = \left(\frac{k}{\bar{r}_n}\right)^{k+1} \frac{r^k}{k!} \exp(-kr/\bar{r}_n) \qquad (2)$$

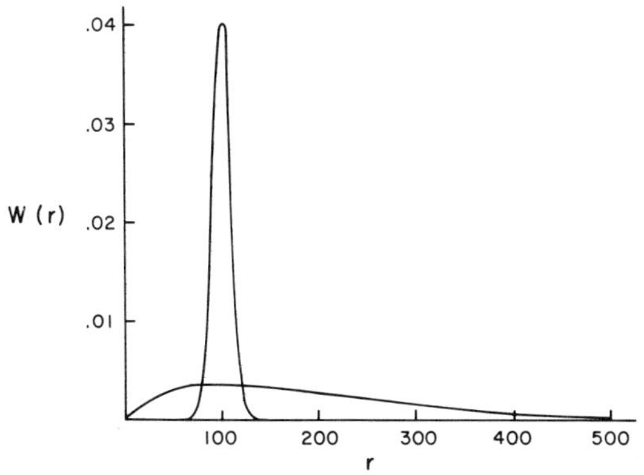

Figure 1.7.1 - Comparison of the Poisson distribution and the Schulz-Flory weight distribution with $\overline{r}_n = 100$. Note the change in scale. (Courtesy J. Am. Chem. Soc.) (Flory, 17).

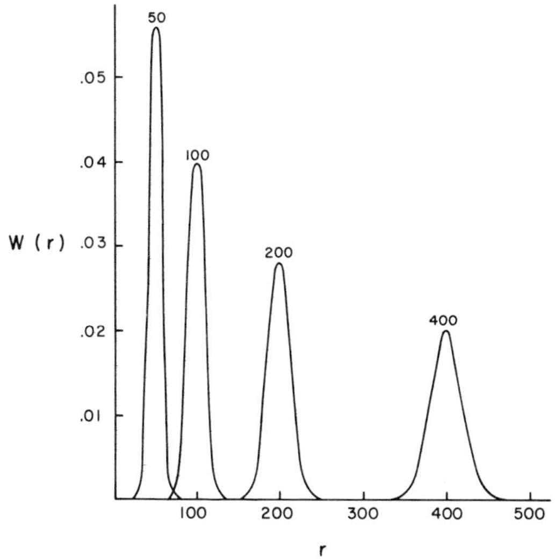

Figure 1.7.2 - Poisson distribution as a function of r for various values of $\overline{r}_n$.

where $\underline{k}$ is a parameter which characterizes the distribution, assumed here to be an integer, and $\bar{r}_n$ is the number average degree of polymerization. If we mix weight fraction w_1 of $\bar{r}_{n1}$ number-average degree of polymerization with weight fraction w_2, $w_1 + w_2 = 1$, of $\bar{r}_{n2}$ number-average degree of polymerization, the combined polymers will have a distribution given by

$$W(r) = w_1 W_1(r) + w_2 W_2(r) \qquad (3)$$

Values of $\underline{r}$ at the maximum and minimum can be found by differentiating equation (3) with respect to $\underline{r}$, setting this equal to zero and simplifying to yield the following equation for $\underline{r}$ when the same value of $\underline{k}$ is used for both subdistributions:

$$\frac{w_2}{w_1}\left(\frac{\bar{r}_{n1}}{\bar{r}_{n2}}\right)^{k+1}\left[1 - \frac{r}{\bar{r}_{n2}}\right]\exp\left[kr\left(\frac{1}{\bar{r}_{n1}} - \frac{1}{\bar{r}_{n2}}\right)\right]$$

$$+\left(1 - \frac{r}{\bar{r}_{n1}}\right) = 0 = f(r) \qquad (4)$$

If $k = 1$ or 2, the distributions which result for either a transfer-type or a recombination-type polymer, respectively, a bimodal distribution will result only when an excess of the high molecular weight material is present: for $k = 1$, $(\bar{r}_1/\bar{r}_2) = 0.1$, (w_2/w_1) must be within the range $2.4 \leqslant w_2/w_1 \leqslant 21.3$. This means that the distribution is very heavily skewed towards the high molecular weight fraction. Table 1.8.1 gives the maximum and minimum values for $\rho = (w_2/w_1)$ as a function of $\bar{r}_2$ for $\bar{r}_{n1} = 100$, k = 1 or 2. Figures 1.8.1 and 1.8.2 show W(r) versus log r for bimodal distributions with k = 1 or 2. Displays such as these are similar to those obtained from fractionation by gel permeation chromatography. The curves appear quite different when drawn in the usual manner, W(r) versus r, Figures 1.8.3 and 1.8.4. Thus whenever bimodal distributions are indicated by fractionation experiments, some care must be taken in the mechanistic interpretation of the results.

TABLE 1.8.1

Conditions for Bimodal Distribution When $\bar{r}_1 = 100$

$(\rho = w_2/w_1)$

	k = 1		k = 2	
$\bar{r}_2$	ρ maximum	ρ minimum	ρ maximum	ρ minimum
200	Unimodal		Unimodal	
300	Unimodal		Unimodal	
400	Unimodal		6.3	5.7
500	Unimodal		8.9	3.0
600	11.8	11.5		
700	13.4	9.3		
800	15.7	6.5	23	0.10
900	18.3	4.0		
1000	21.3	2.4	39	0.006

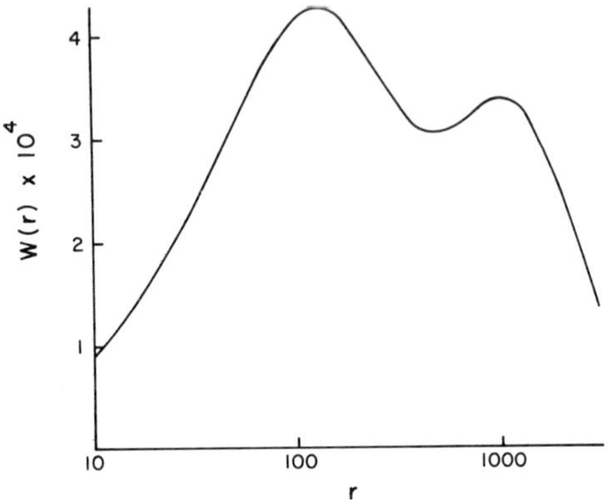

Figure 1.8.1 - Weight distribution for a bimodal polymer made from two "most probable" distributed polymers with 9.09 weight % $\bar{r}_1 = 100$ and 90.91 weight % $\bar{r}_2 = 1000$ as a function of log r.

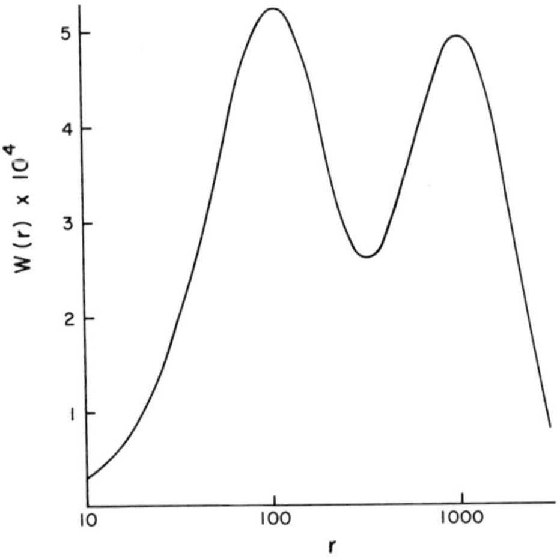

Figure 1.8.2 - Weight distribution for a bimodal polymer made from two Schulz-type polymers with k = 2, 9.09 weight % $\bar{r}_1 = 100$ and 90.91 weight % $\bar{r}_2 = 1000$ as a function of log r.

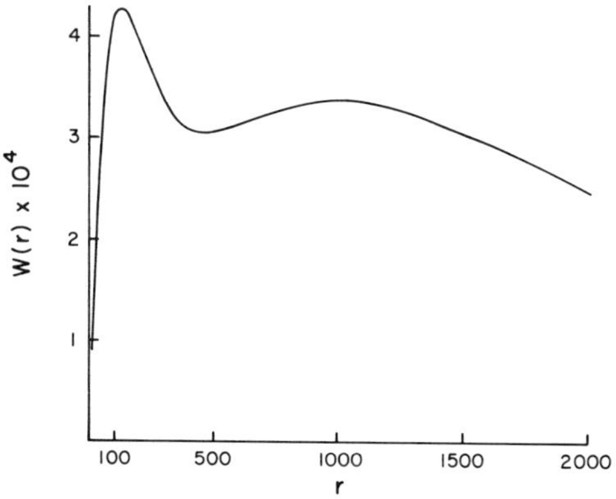

Figure 1.8.3 - Weight distribution for a bimodal polymer as a function of r. Same conditions as in Figure 1.8.1.

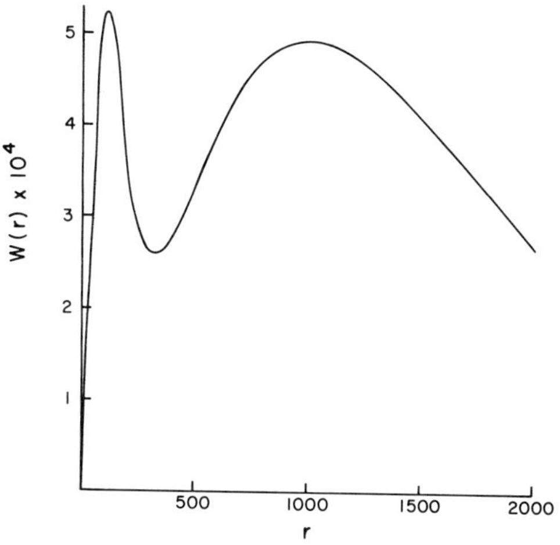

Figure 1.8.4 - Weight distribution for a bimodal polymer as a function of r. Same conditions as in Figure 1.8.2.

9. Some Useful Sum Terms

The molecular weight distribution of a polymer is usually derived in one of two ways. Either the differential equations for the concentration of each species, based on an assumed mechanism of polymerization, is solved for the concentration as a function of some reduced time variable or the probabilities of finding each species as a function of the extent of reaction is determined by a statistical approach. In either case we usually obtain the concentration of each species as a function of other parameters. One means of obtaining the normalized functions F(r) and W(r) as well as the average degrees of polymerization, $\bar{r}_n$, $\bar{r}_w$, $\bar{r}_z$, and so on, is to sum the individual equations. Alternately, the differential equation for the rate of change of the species with time, dP_r/dt, can be summed directly, that is, finding $d(\Sigma r^i P_r)/dt$, which may result in an equation which is easily solved. The list below is a useful collection of various sums.

$$\sum_{x=1}^{\infty} p^{x-1} = 1/(1 - p) \qquad p < 1$$

$$\sum_{x=1}^{\infty} xp^{x-1} = 1/(1 - p)^2 \qquad p < 1$$

$$\sum_{x=1}^{\infty} x^2 p^{x-1} = (1 + p)/(1 - p)^3 \qquad p < 1$$

$$\sum_{x=1}^{\infty} x^n p^{x-1} = \frac{d}{dp} [p \sum_{x=1}^{\infty} x^{n-1} p^{x-1}] \qquad p < 1$$

$$\sum_{i=0}^{n} a^{n-i} b^i n!/(n - i)!i! = (a + b)^n \qquad \text{n is an integer}$$

$$\sum_{i=0}^{n} a^i i n!/(n - i)!i! = na(1 + a)^{n-1}$$

$$\sum_{i=0}^{n} a^i i^2 n!/(n - i)!i! = na(1 + na)(1 + a)^{n-2}$$

$$\sum_{x=1}^{\infty} A^{x-1}/(s - 1)! = \exp A$$

$$\sum_{x=1}^{\infty} xA^{x-1}/(x - 1)! = (1 + A) \exp A$$

$$\sum_{x=1}^{\infty} x^2 A^{x-1}/(x - 1)! = (1 + 3A + A^2) \exp A$$

$$\sum_{x=1}^{\infty} x^n A^{x-1}/(x - 1)! = \frac{d}{dA} [A \sum_{x=1}^{\infty} x^{n-1} A^{x-1}/(x - 1)!]$$

$$\sum_{x=2}^{\infty} \sum_{s=1}^{x-1} N_s N_{x-s} = (\sum_{s=1}^{\infty} N_s)^2$$

$$\sum_{x=2}^{\infty} \sum_{s=1}^{x-1} x N_s N_{x-s} = 2\left(\sum_{s=1}^{\infty} N_s\right)\left(\sum_{s=1}^{\infty} s N_s\right)$$

$$\sum_{x=2}^{\infty} \sum_{s=1}^{x-1} x^2 N_s N_{x-s} = 2\left(\sum_{s=1}^{\infty} s N_s\right)^2 + 2\left(\sum_{s=1}^{\infty} N_s\right)\left(\sum_{s=1}^{\infty} s^2 N_s\right)$$

10. Practical Methods of Establishing Molecular Weight Distributions

Two general methods are used for determining the distribution of molecular weights in a polymer: by determination of various averages of the distribution and by fractionation.

The first method requires additional assumptions because two or three averages of a distribution really cannot define a material that contains perhaps hundreds or thousands of distinguishable entities (distinguishable in theory, but not in practice). The presence of comonomers, branches, stereoplacements, defects, (polymethacrylonitrile may contain ketenimine linkages), and so on, further complicate the situation. Despite these complications, measurements of the various molecular weight averages are useful.

A good review of number-average methods is found in Number Average Molecular Weights (18) which was published before the advent of the high-speed membrane osmometer for the medium to high molecular weight range and the vapor phase osmometer for the low to medium range (19).

These two instruments give very rapid results (in terms of minutes) in contrast to the long periods of time sometimes required for the more classical methods (weeks).

The classical viscosity methods either in dilute solution (specific or intrinsic viscosity) or in the melt are rapid and simple, but require calibration by absolute methods. They are extremely useful for routine measurements, but care must be exercised if different amounts of branching, composition (as in a comonomer), degradation, or other chemical changes have occurred, since these may alter the calibration.

The weight-average molecular weights can be determined by ultracentrifugation and diffusion measurements and by light scattering measurements. The latter measurements are fairly easy in the commercial instruments available provided that care is used to remove extraneous scattering centers from the solutions (dust removal). Determination of $\overline{r}_w$ by ultracentrifugation and diffusion measurements requires specialized equipment and techniques; however the z-average molecular weight can be obtained by this method. Density gradient sedimentation also provides measures of the average molecular weights (20).

The methods enumerated above are generally applicable to any soluble polymer; details of the methods are to be found in many texts (21-32). Cheung reviews methods used for insoluble polymers (33). For any given polymer there probably exist other methods that are equally effective measures of an average molecular weight. Wolf gives an extended list of properties that vary with molecular weight (34). As an instance, the number-average molecular weight of undegraded commercial Nylon 66 is easily measured by determining the total amine, carboxyl, and acetyl (capping agent) contents, but degradation by heat or light will form inactive ends. This particular chemical method will not account for the small amount of cyclic oligomers present in the polymer.

In Polymer Fractionation (35) some ten different methods are reviewed in detail: fractional precipitation, fractional solution, fractional chromatography, gel permeation chromatography, thermal and isothermal diffusion, sedimentation, turbidimetric titration, summative fractionation, and rheological methods; there is also a chapter on less well developed methods. Guzman has provided a detailed list of systems used in fractional precipitation (36). One of the most rapid methods of fractionation is gel permeation chromatography (GPC) (37, 38) which has been the subject of several recent symposia (19, 39-41). The method does require calibration and is subject to errors of interpretation when branching, compositional, or other variations occur from sample to sample. Details of analysis of the GPC curves are given elsewhere (37, 39, 41).

To determine whether a given polymeric system conforms to a specified molecular weight distribution is not an easy task; it involves repeated measurements of either the molecular weight distributions or the average molecular weights of a series of polymers prepared under various conditions. Hopefully the figures presented in this book will aid the researcher in visualizing how the various parameters influence the molecular weight distributions.

11. References

1. G. V. Schulz, "Uber die Beziehung zwischen Reaktionsgeschwindigkeit und Zusammensetzung des Reaktionsproduktes bei Makropolymerisationsvorgängen," Z. Physik. Chem., B30, 379 (1935).
2. P. J. Flory, "Molecular Size Distribution in Linear Condensation Polymers," J. Am. Chem. Soc., 58, 1877 (1936).
3. P. J. Flory, Principles of Polymer Chemistry, Cornell University Press, Ithaca, N. Y., 1953.
4. A. E. Fainerman and V. M. Polyakova, "About a Method of Linearization of the Most Probable Flory's Distribution," Vysokomol. Soedin., A10(5), 1214 (1958).
5. G. V. Schulz, "Uber die Kinetik der Kettenpolymerisationen. V. Der Einfluss verschiedener Reaktionsarten auf die Polymolekularität," Z. Physik. Chem., B43, 25 (1939).
6. B. H. Zimm, "Apparatus and Methods for Measurement and Interpretation of the Angular Variation of Light Scattering; Preliminary Results on Polystyrene Solutions," J. Chem. Phys., 16, 1099 (1948).

7. A. M. Kotliar, "Critical Analysis of Molecular Weight Distributions Derived from Fractionation Data. I. Column Elution," J. Polymer Sci., A2, 1373 (1964).
8. L. T. Muus and W. H. Stockmayer, quoted by F. W. Billmeyer, Jr., in Textbook of Polymer Chemistry, Interscience, New York-London, 1962.
9. W. Weibull, "A Statistical Distribution Function of Wide Applicability," J. Appl. Mech., 18, 293 (1951).
10. L. H. Tung, "Fractionation of Polyethylene," J. Polymer Sci., 20, 495 (1956).
11. H. Wesslau, "Die Molekulargewichtsverteilung einiger Niederdruckpolyäthylene," Makromol. Chem., 20, 111 (1956).
12. W. F. Espenscheid, M. Kerker, and E. Matijević, "Logarithmic Distribution Functions for Colloidal Particles," J. Phys. Chem., 68, 3093 (1964).
13. E. P. Honig, "Logarithmic Distribution Functions for Colloidal Particles," J. Phys. Chem., 69, 4418 (1965).
14. W. D. Lansing and E. O. Kramer, "Molecular Weight Analysis of Mixtures by Sedimentation Equilibrium in the Svedberg Ultracentrifuge," J. Am. Chem. Soc., 57 1369 (1935).
15. A. M. Kotliar, "A Critical Evaluation of Mathematical Molecular Weight Distribution Models Proposed for Real Polymer Distributions. I. Effects of a Low Molecular Weight Cut-Off Value," J. Polymer Sci., A2, 4303 (1964).
16. A. M. Kotliar, "A Critical Evaluation of Mathematical Molecular Weight Distribution Models Proposed for

Real Polymer Distributions. II. Effects of a High Molecular Weight Cut-Off Value," J. Polymer Sci., A2, 4327 (1964).

17. P. J. Flory, "Molecular Size Distribution in Ethylene Oxide Polymers," J. Am. Chem. Soc., 62, 1561 (1940).

18. R. U. Bonner, M. Dimbat, and F. H. Stross, Number Average Molecular Weights, Interscience, New York, 1958.

19. J. Mitchell and F. W. Billmeyer, Analysis and Fractionation of Polymers, Symposium, Chicago, 1964, Wiley, New York, 1965.

20. J. J. Hermans and H. A. Ende, "Density Gradient Centrifugation of a Polymer-Homologous Mixture," J. Polymer Sci., C1, 161 (1963); J. J. Hermans, "Density Gradient Centrifugation of a Mixture of Polymers Differing in Molecular Weight and Specific Volume," J. Polymer Sci., C1, 179 (1963); H. A. Ende, "Molecular Weight Averages as Determined by Density Gradient Centrifugation," J. Polymer Sci., B3, 139 (1965).

21. K. A. Stacey, Light Scattering in Physical Chemistry, Butterworths, London, 1956.

22. S. R. Rafikov, Determination of Molecular Weights and Polydispersity of High Polymers, Israel Program for Sci. Trans., 1964.

23. H. A. Stuart, Die Physik Der Hochpolymeren. Volume II. Das Makromolekül in Lösungen, Springer-Verlag, Berlin, 1953.

24. J. K. Stille, Introduction to Polymer Chemistry, Wiley, New York, 1962.

25. A. Weissberger, Physical Methods of Organic Chemistry, 3rd Edition, Parts I-IV, Interscience, New York, 1959.
26. D. Margerison and G. C. East, An Introduction to Polymer Chemistry, Pergamon, London, 1967.
27. D. A. Smith, Addition Polymers: Formation and Characterization, Plenum Press, New York, 1968.
28. P. W. Allen, Techniques of Polymer Characterization, Butterworths, London, 1959.
29. F. W. Billmeyer, Textbook of Polymer Chemistry, Interscience, New York, 1962.
30. D. McIntyre and F. Gornick, Light Scattering from Dilute Polymer Solutions, Collection, Gordon and Breach, New York, 1964.
31. H. Morawetz, Macromolecules in Solution, Interscience, New York, 1965.
32. D. McIntyre, Characterization of Macromolecular Structure, Conference, Warrenton, Va., 1967, National Academy of Sciences, Washington, D. C., 1968.
33. H. C. Cheung, "Insoluble Polymers: Molecular Weights and Their Distributions," in Physical Methods in Macromolecular Chemistry, Vol. I., B. Carroll, Editor, Dekker, 1969.
34. K. A. Wolf, Struktur und physikalisches Verhalten der Kunststoffe. Springer-Verlag, Berlin, 1962.
35. M. J. R. Cantow, Polymer Fractionation, Academic Press, New York, 1967.
36. G. M. Guzman, "Fractionation of Polymers," in Polymer Handbook, J. Brandrup and E. H. Immergut, Editors, Interscience, New York, 1966.

37. J. F. Johnson, R. S. Porter and M. J. R. Cantow, "Gel Permeation Chromatography with Organic Solvents," Rev. Macromol. Chem., 1, 393 (1966).

38. H. Determann, "Chromatographic Separations on Porous Gels," Angew. Chem., Int. Ed., 3, 608 (1964).

39. K. A. Boni and F. A. Sliemers, International Symposium on Polymer Characterization, Interscience, New York, 1969.

40. J. F. Johnson and R. S. Porter, Analytical Gel Permeation Chromatography, Symposium, Chicago, 1967, Wiley, New York, 1968.

41. ACS Meeting, Houston, February, 1970.

Chapter 2

Addition Polymerization with Termination

Contents

1. Introduction

In this chapter we examine the distribution functions that result from addition polymerization in which the active species disappear either by a first-order or a second-order termination reaction. We make no distinction between free radical, cationic, anionic, or coordination type polymerization. However, free radicals are usually destroyed by a second-order mechanism; hence the distribution functions are applicable to polymerization by free-radical means. All active species are represented by an asterisk (*) and concentrations are expressed as moles per liter.

The initiation of polymerization may occur in a variety of ways:

1. Initiation may be considered to be constant over the entire time of polymerization. Here the rate of initiation is expressed as R_I. The term R_I in the equation may be replaced by some function of the light intensity, the rate of production of active species (that is, $R_I = 2k_d$ [initiator] if two active fragments are produced by the first-order decomposition constant, k_d, for the initiator) or by an equation which involves the monomer concentration if the monomer concentration is considered to be held constant (such may be the case in thermal initiation).

2. Both ends of the growing chain may be active, in which case R_I refers to the production of active ends, not active molecules.

3. Thermal initiation may require bimolecular monomer initiation, thus $R_I = k_i M^2$.

4. The active initiator may react slowly with monomer $R_I = k_i MI^*$.

5. In some cases the initiator reacts wastefully or becomes deactivated during the polymerization time.

The concentration of monomer may be considered as held constant throughout the polymerization either by stopping the reaction before any appreciable amount of monomer has been used or by continually adding monomer during polymerization. The formulas for this case are much simpler than when the monomer concentration is allowed to vary.

Propagation is assumed to be independent of chain length when the chains are large. In some cases, the reaction between the initiator fragment, I*, and monomer may have a different rate than the reaction between polymer fragment, R_r^*, and monomer

$$I^* + M \rightarrow R_1^* \qquad k_i \text{ or } k_p \tag{1}$$

$$R_r^* + M \rightarrow R_{r+1}^* \qquad k_p \tag{2}$$

Various transfer reactions may occur: transfer to monomer

$$R_r^* + M \rightarrow P_r + R_1^* \qquad k_{tr,m} \tag{3}$$

transfer to solvent

$$R_r^* + S \rightarrow P_r + S^* \qquad k_{tr,s} \tag{4}$$

and transfer to polymer

$$R_r^* + P_s \rightarrow P_r + R_s^* \qquad k_{tr,p} \tag{5}$$

A source of confusion has arisen over the transfer-to-initiator reaction. We neglect the second-order reaction

$$R_r^* + \text{initiator} \rightarrow P_r + \text{initiator}^* \tag{6}$$

which may occur as an induced decomposition of the initiator or may produce an initiator fragment that is still capable of decomposition into active species at a later time. The extension of the present formulas to include these effects is straightforward. However, in ionic polymerization, an active fragment can be expelled from a growing center, forming polymer and an active initiator fragment by a first-order mechanism:

$$R_r^* \rightarrow P_r + I^* \qquad k_{iex} \tag{7}$$

This is called here the "initiator expulsion reaction."

Termination of growing chains can occur in a variety of forms. Of course the "living polymers" do not have any termination and are considered in Chapter 3.

Active centers can terminate spontaneously by a first-order reaction, which is sometimes called catalyst deactivation,

$$R_r^* \rightarrow P_r \qquad k_t \tag{8}$$

or can interact by a second-order mechanism with monomer,

$$R_r^* + M \rightarrow P_{r+1} \text{ (or } P_r + M) \qquad k_{t,m} \tag{9}$$

or two active centers can react either by combination

$$R_r^* + R_s^* \rightarrow P_{r+s} \qquad k_{t,c} \tag{10}$$

or by disproportionation, usually by transfer of an atom or a group from one molecule to another

$$R_r^* + R_s^* \rightarrow P_r + P_s \qquad k_{t,d} \tag{11}$$

The last two reactions are the primary means of termination to be considered in this chapter.

When transfer to solvent occurs

$$R_r^* + S \rightarrow P_r + S^* \qquad k_{tr,s} \tag{12}$$

the S* fragment can disappear in at least three ways: by solvent reinitiation,

$$S^* + M \rightarrow R_1^* \qquad k_{s,i} \tag{13}$$

solvent dimerization,

$$2S^* \rightarrow \text{products} \qquad k_{s,s} \tag{14}$$

and solvent termination,

$$S* + R_r* \rightarrow P_r \qquad k_{t,s} \tag{15}$$

For long chains it is unimportant whether solvent termination occurs by combination or by disproportionation.

In most cases, the assumption of steady state kinetics is made wherein the rate of formation of active fragments equals their rate of disappearance.

$$\frac{-dR*}{dt} = R_I - 2k_t R*^2 \tag{16}$$

In this equation, we are not concerned with whether termination is by disproportionation or by combination but rather that termination is a second-order reaction and that two active radicals are destroyed by termination. If R_I is a constant, then equation (16) is easily integrated:

$$R* = \left(\frac{R_I}{2k_t}\right)^{\frac{1}{2}} \tanh\ (2k_t t) \tag{17}$$

Figure 2.1.1 shows tanh x versus x. After $x = 2k_t t = 3$, the value of tanh x is essentially unity [tanh 3 = 0.99505], so that we obtain the same value of R* as if we set equation (16) to zero and solved for R* directly. Davis (1) and Bamford (2) have considered the more complicated kinetic system

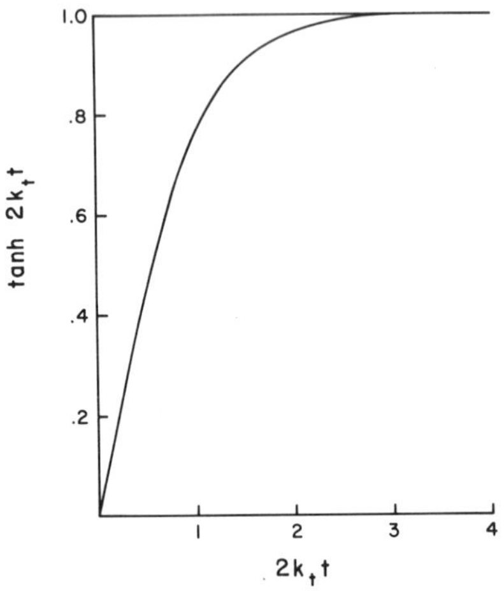

Figure 2.1.1 - Equation 2.1.17 relates the active species concentration R* to the reduced time variable $2k_t t$. When this variable is equal to or greater than 3, R* is independent of time.

$$I \xrightarrow{k_d} R_1^*$$

$$R_n^* + M \xrightarrow{k_p} R_{n+1}^* \qquad (18)$$

$$R_r^* + R_s^* \xrightarrow{k_t} \text{polymers}$$

where the generation of radicals is

$$R_I = k_d I = k_d I_o \exp(-k_d t) \qquad (19)$$

The result is

$$R^* = [k_d I_o e^{-k_d t}/2k_t]^{\frac{1}{2}} \left[\frac{\tilde{I}_1(a)\tilde{K}_1(ax) - \tilde{K}_1(a)\tilde{I}_1(ax)}{\tilde{I}_1(a)\tilde{K}_0(ax) + \tilde{K}_1(a)\tilde{I}_0(ax)}\right] \quad (20)$$

(Davis, 1; Bamford, 2) where $a = 2(2k_t I_o/k_d)^{\frac{1}{2}}$

$$x = \exp(-k_d t/2)$$

and $\tilde{I}_0$, $\tilde{K}_0$, $\tilde{I}_1$, $\tilde{K}_1$ are the modified Bessel functions of the zero and first order respectively.

If the half-life of the initiator I is much larger than the lifetime of radicals, the concentration of radicals will be a slowly decreasing function of time. As a first approximation, we can use only the square root term in equation (20); this also results from replacing R_I in equation (16) with equation (19):

$$R^* \approx [k_d I_o \exp(-k_d t)/2k_t]^{\frac{1}{2}} \quad (21)$$

Chien (3) has used equation (20) to derive the average molecular weights without use of the steady state assumption (Section 8). Inspection of these equations quickly indicates why the exact solution is not used more often!

In the derivations, the assumption is made that long chains are produced. Thus the rate of disappearance of monomer is written as

$$-dM/dt = k_p R^* M$$

instead of the more exact expression

$$-dM/dt = k_i MI^* + k_p R^*M + k_{tr,m} R^*M$$

when initiation occurs by reaction (1), propagation by reaction (2) and transfer to monomer by reaction (3). Neglect of the first and third terms relative to the second term becomes important only when low molecular weight polymer is produced. Thus if application of the rate and distribution equations is to be made to low molecular weight polymer, it is best to recheck the derivations.

2. Invariant Monomer Concentration

A. Constant rate of initiation, monomer concentration invariant, no transfer, termination by second-order combination (4-7).

The distribution is the Schulz distribution with k = 2 (Figures 1.3.1 and 1.3.2, pages 13 and 14), which can be written as

$$W(r) = (-\ln p)^3 r^2 p^r/2 \tag{1}$$

$$= (4r^2/\bar{r}_n^{\,3}) \exp(-2r/\bar{r}_n) \tag{2}$$

(Schulz, 7). In equation (2),

$$\int_0^\infty W(r)\, dr = 1 \tag{3}$$

with the approximation that $-\ln p = 1 - p$, which is valid for high molecular weight polymer, that is, $p \approx 1$; then

$\sum_{r=0}^{\infty} W(r) = 1$. The number average degree of polymerization is

$$\bar{r}_n = 2(k_p M + \{2R_I\, k_{t,c}\}^{\frac{1}{2}})/\{2R_I\, k_{t,c}\}^{\frac{1}{2}} \qquad (4)$$

$$= 2/(-\ln p)$$

If $\{2R_I k_{t,c}\}^{\frac{1}{2}} << k_p M$, reasonable for high molecular weight polymer, then

$$\bar{r}_n/2 = \bar{r}_w/3 = k_p M/\{2R_I k_{t,c}\}^{\frac{1}{2}} \qquad (5)$$

Thus the ratio $\bar{r}_w/\bar{r}_n = 1.5$

B. Constant rate of initiation, monomer concentration invariant, no transfer, termination by second-order disproportionation (4-6).

The distribution is the Schulz distribution with k = 1, the Schulz-Flory most probable distribution:

$$W(r) = r(1 - \zeta)^2 \zeta^{r-1} \qquad \sum_{r=1}^{\infty} W(r) = 1$$

$$= (r/\bar{r}_n^{\,2}) \exp(-r/\bar{r}_n) \qquad \int_0^{\infty} W(r)\, dr = 1 \qquad (6)$$

where

$$\zeta = k_p M/[k_p M + \{2\, R_I k_{t,d}\}^{\frac{1}{2}}] \qquad (7)$$

$$\bar{r}_n = 1/(1 - \zeta) \qquad (8)$$

$$\bar{r}_w = (1 + \zeta)/(1 - \zeta) \tag{9}$$

Since for high molecular weight polymers $\zeta \approx 1$, the ratio

$$\bar{r}_w/\bar{r}_n = 2.0 \tag{10}$$

The distributions are shown in Figures 1.2.1 and 1.2.2, page 10 and 11.

C. Constant rate of initiation, monomer concentration invariant, transfer to monomer and to solvent, termination by second-order disproportionation (5, 6).

Addition of the transfer-to-monomer reaction and the transfer-to-solvent reaction to the kinetic scheme has no influence on the shape of the distribution. The equations of Section 2B are valid with the exception that

$$\zeta = k_p M/[k_p M + k_{tr,m} M + k_{tr,s} S + \{2R_I\, k_{t,d}\}^{\frac{1}{2}}] \tag{11}$$

D. Constant rate of initiation, monomer concentration invariant, transfer to monomer and to solvent, termination by second-order combination.

Addition of the transfer-to-monomer reaction and the transfer-to-solvent reaction to the kinetic scheme cause changes in the distribution, depending upon the importance of the transfer reactions when termination is solely by second-order combination. We now wish to introduce the reduced variable γ. In this section

$$\gamma = \frac{k_{tr,m}M + k_{tr,s}S}{\{2R_I k_{t,c}\}^{\frac{1}{2}}} \tag{12}$$

It will take other definitions in other sections.

$$W(r) = \frac{r(1 - \zeta)^2 \zeta^{r-3}}{2(1 + \gamma)} \{2\gamma\zeta + (r - 1)(1 - \zeta)\} \tag{13}$$

(Bamford et al., 5) with

$$\zeta = k_p M/[k_p M + k_{tr,m}M + k_{tr,s}S + \{2R_I k_{t,c}\}^{\frac{1}{2}}]$$

The distributions F(r) and W(r) are shown in Figures 2.2.1 and 2.2.2. When $\gamma = 0$, this reduces to the equations of Section 2A (for long chains). As γ increases, the distribution function changes shape and approaches the Schulz-Flory distribution as γ becomes large. Furthermore, the maximum in the frequency distribution (Figure 2.2.1) gradually disappears.

The average molecular weights are

$$\bar{r}_n = 2k_p M/\{2R_I k_{t,c}\}^{\frac{1}{2}}(1 + 2\gamma) \tag{14}$$

$$= 2(1 + \gamma)/(1 + 2\gamma)(1 - \zeta) \tag{15}$$

$$= 2k_p M/[2k_{tr,m}M + 2k_{tr,s}S + \{2R_I k_{t,c}\}^{\frac{1}{2}}] \tag{16}$$

The last equation is sometimes written as

Curve a, termination by disproportionation.

Curve	b	c	d	e
γ	0	0.05	0.25	0.50

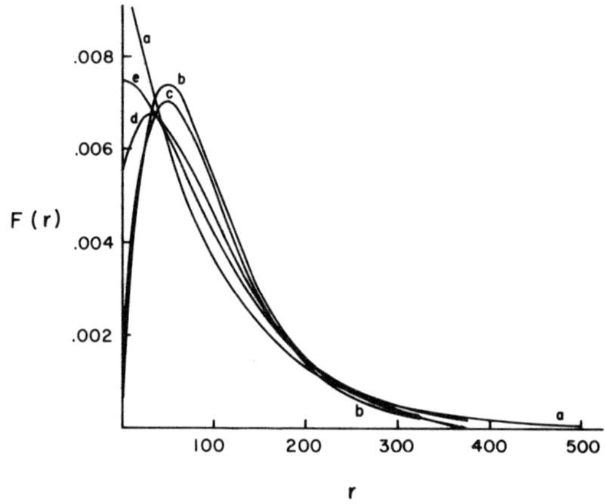

Figure 2.2.1 - The frequency or mole fraction distribution for constant rate of initiation, monomer concentration invariant, transfer to monomer and to solvent, termination by second-order combination $\bar{r}_n = 100$, $\gamma = (k_{tr,m}M + k_{tr,s}S)/[2R_I k_{t,c}]^{\frac{1}{2}}$ varies. The low molecular weight tail becomes progressively more important as γ increases. (After Bamford et al., 5)

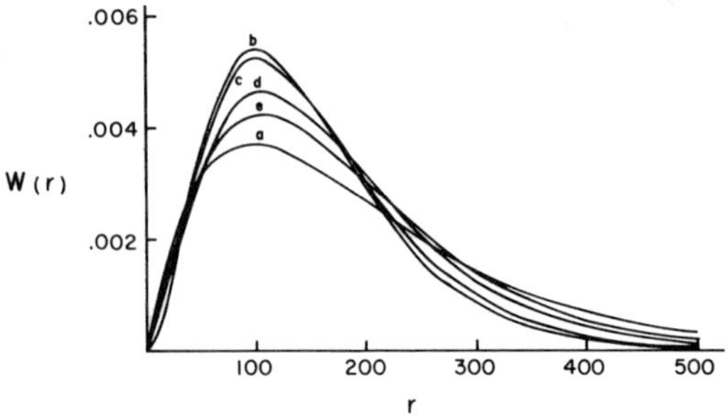

Figure 2.2.2 - Weight fraction distribution for the same

conditions as in Figure 2.2.1.

$$1/\bar{r}_n = 1/DP = C_m + C_s \, S/M + \{2R_I k_{t,c}\}^{\frac{1}{2}}/2k_p M \tag{17}$$

where

$$C_m = k_{tr,m}/k_p, \quad C_s = k_{tr,s}/k_p \tag{18}$$

$$\bar{r}_w = k_p M(2\gamma + 3)/\{2R_I k_{t,c}\}^{\frac{1}{2}}(\gamma + 1)^2 \tag{19}$$

The ratio $\bar{r}_w/\bar{r}_n$ varies from 1.5 when $\gamma = 0$ to 2.0 when γ is large.

$$\bar{r}_w/\bar{r}_n = (3 + 2\gamma)(1 + 2\gamma)/2(1 + \gamma)^2 \tag{20}$$

This equation is shown in Figure 2.2.3.

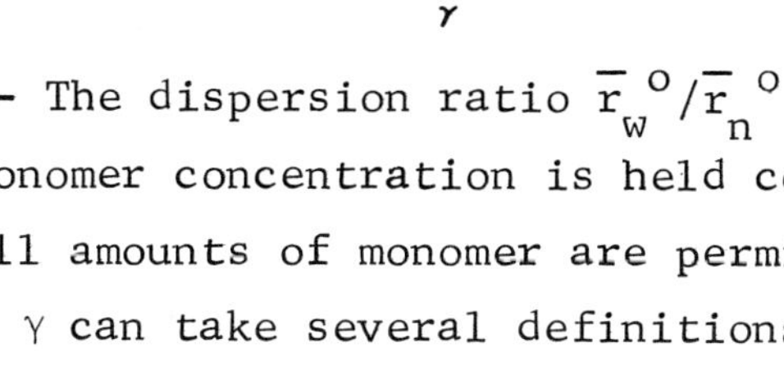

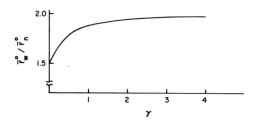

Figure 2.2.3 - The dispersion ratio $\bar{r}_w^{\,o}/\bar{r}_n^{\,o}$ is calculated either when monomer concentration is held constant or when extremely small amounts of monomer are permitted to react. The parameter γ can take several definitions: see sets 2D, E, and F. In general

$$\gamma = (1 + k_{t,d}/k_{t,c})^{\frac{1}{2}} \frac{[k_{tr,m}M + k_{tr,s}S]}{[2R_I k_{t,c}]^{\frac{1}{2}}} + \frac{k_{t,d}}{k_{t,c}}$$

From equation (13) the fraction of molecules formed by transfer, $\Omega(r)$, can be determined as r varies across the distribution. Combination of (13) with (15) yields

$$\Omega(r) = 1/[1 + (r - 1)\ (1 + \gamma)/\bar{r}_n \gamma\varsigma(1 + 2\gamma) \qquad (21)$$

which is plotted in Figure 2.2.4 for $\bar{r}_n = 100$. This would be useful if one could differentiate between molecules formed by combination and those formed by transfer. The former would contain two initiator fragments, whereas those formed by transfer would contain either one or no initiator fragments; the other end group would be the product of the transfer reaction.

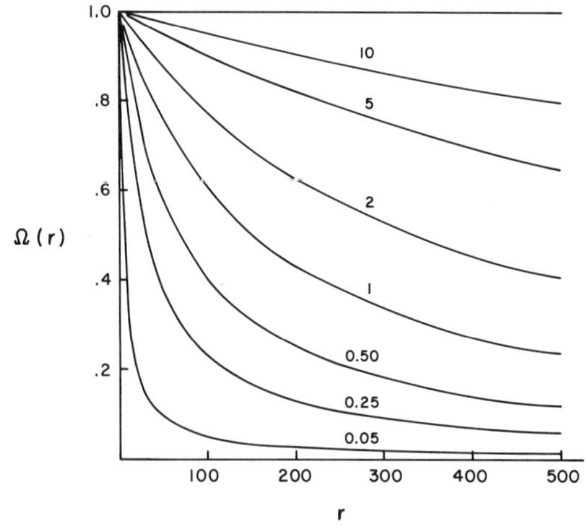

Figure 2.2.4 - Fraction of molecules formed by transfer when termination is by second-order combination, $\bar{r}_n = 100$, M constant, as a function of r for various values of γ.

E. Constant rate of initiation, monomer concentration invariant, no transfer, termination by second-order combination and disproportionation (5).

Here we wish to examine the distribution which results when second-order termination can occur either by combination or disproportionation without the added complication of the transfer reactions. Let us redefine the parameter γ such that

$$\gamma = k_{t,d}/k_{t,c} \tag{21}$$

By so doing we obtain the same distribution described in Section 2D and shown in Figures 2.2.1 and 2.2.2 with the exception that

$$\zeta = k_pM/(k_pM + \{2R_I\ (k_{t,c} + k_{t,d})\}^{\frac{1}{2}}) \tag{22}$$

The average molecular weights are

$$\bar{r}_n = 2k_pM(1 + \gamma)^{\frac{1}{2}}/\{2R_Ik_{t,c}\}^{\frac{1}{2}}\ (1 + 2\gamma) \tag{23}$$

$$\bar{r}_w = k_pM\ (3 + 2\gamma)/\{2R_Ik_{t,c}\}^{\frac{1}{2}}\ (1 + \gamma)^{3/2} \tag{24}$$

which are not the same as equations (2.14) and (2.19). However the ratio $\bar{r}_w/\bar{r}_n$ is the same as equation (2.20) and as shown in Figure 2.2.3.

F. Constant rate of initiation, monomer concentration invariant, transfer to monomer and to solvent, termination by second-order combination and disproportionation.

In the preceding five sets, we have separated the assorted transfer and termination reactions so as to be able to examine their influence on the distribution. Here we present the combined distribution functions:

$$W(r) = \frac{r(1 - \zeta)\zeta^{r-1}}{k_p M} \Bigg\{ k_{tr,m}M + k_{tr,s}S$$

$$+ \left[\frac{R_I}{2(k_{t,c} + k_{t,d})}\right]^{\frac{1}{2}} \left[\frac{2(r - 1)k_{t,c}}{2k_p M} [k_{tr,m}M + k_{tr,s}S\right.$$

$$\left. + \{2R_I(k_{t,c} + k_{t,d})\}^{\frac{1}{2}}] + 2k_{t,d}\right]\Bigg\} \tag{25}$$

(Bamford et al., 5) where

$$\zeta = k_p M/[k_p M + k_{tr,m}M + k_{tr,s}S$$

$$+ \{2R_I(k_{t,c} + k_{t,d})\}^{\frac{1}{2}}]$$

If we defined

$$\gamma = (1 + k_{t,d}/k_{t,c})^{\frac{1}{2}} \frac{[k_{tr,m}M + k_{tr,s}S]}{\{2R_I k_{t,c}\}^{\frac{1}{2}}} + \frac{k_{t,d}}{k_{t,c}} \tag{26}$$

we again would obtain the distribution in Section 2D and shown in Figures 2.2.1 and 2.2.2.

The average molecular weights are

$$\bar{r}_n = k_p M \Bigg/ \left\{ k_{tr,m}M + k_{tr,s}S + \left[\frac{R_I}{2(k_{t,c} + k_{t,d})}\right]^{\frac{1}{2}} (k_{t,c} + 2k_{t,d}) \right\} \tag{27}$$

$$\bar{r}_w = 2k_p M[k_{tr,m}M + k_{tr,s}S + \{R_I/2(k_{t,c} + k_{t,d})\}^{\frac{1}{2}}(3k_{t,c} + 2k_{t,d})] \div [k_{tr,m}M + k_{tr,s}S + \{2R_I(k_{t,c} + k_{t,d})\}^{\frac{1}{2}}]^2 \tag{28}$$

The ratio $\bar{r}_w/\bar{r}_n$ is given in Figure 2.2.3.

3. Monomer Concentration Varies, No Transfer-to-Monomer Reaction

A. Constant rate of initiation, monomer concentration varies, no transfer, termination by second-order combination.

We now look at what happens to the distribution when the monomer concentration is allowed to vary. We consider two cases:

1. The initial conditions are such that the polymer first formed has a number-average degree of polymerization of 100, $\bar{r}_n^{\,o} = 100$; thus $\bar{r}_n$ will vary with conversion.

2. The initial conditions are adjusted so that

$\bar{r}_n = 100$ at the conversion being examined; thus $\bar{r}_n^{\,o}$ is different for each conversion.

When the polymer molecules are formed only by second-order termination by combination, the distribution is

$$W(r) = \frac{r\mu}{M_o - M}\left[\left\{\frac{1}{r-1} - \frac{r-3}{2(r-1)(r-2)}\right\}\right.$$

$$\times \left\{\left(1 + \frac{\mu}{M_o}\right)^{1-r} - \left(1 + \frac{\mu}{M}\right)^{1-r}\right\}$$

$$\left. + \frac{\mu(r-1)}{2(r-2)}\left\{\frac{1}{M_o}\left(1 + \frac{\mu}{M_o}\right)^{1-r} - \frac{1}{M}\left(1 + \frac{\mu}{M}\right)^{1-r}\right\}\right] \tag{1}$$

(Bamford et al., 5) where

$$\mu = \{2R_I k_{t,c}\}^{\frac{1}{2}}/k_p \tag{2}$$

The curves are given in Figures 2.3.1 and 2.3.2 for $\bar{r}_n^{\,o} = 100$. As the conversion increases, $\bar{r}_n$ decreases, and the maximum of distribution shifts toward the lower values of <u>r</u>. The breadth of the distribution becomes broader because $\bar{r}_w$ does not decrease as rapidly as $\bar{r}_n$:

$$\bar{r}_n = 2(M_o - M)\, k_p/[\ln(M_o/M)]\{2R_I k_{t,c}\}^{\frac{1}{2}} \tag{3}$$

$$\bar{r}_w = 3(M_o + M)\, k_p/2\{2R_I k_{t,c}\}^{\frac{1}{2}} \tag{4}$$

Curve	a	b	c	d	e	f
c	0.01	0.10	0.25	0.50	0.75	0.90
$\bar{r}_n$	99.5	94.8	87.0	72.0	54.1	39.1

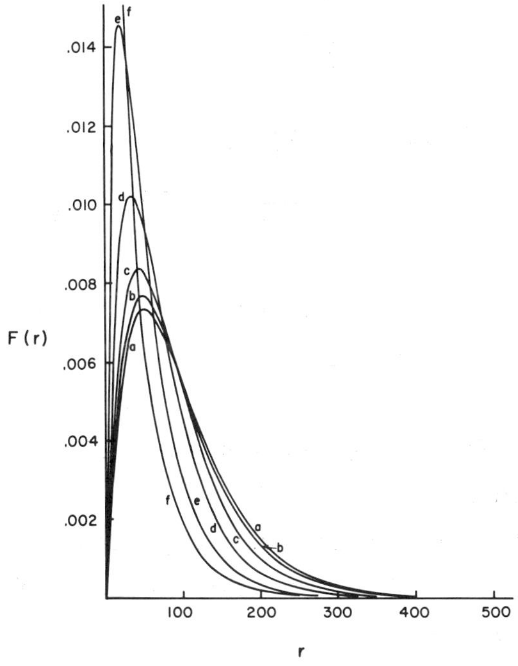

Figure 2.3.1 - Frequency distribution as a function of $\underline{r}$ for constant rate of initiation, monomer concentration varies, no transfer, termination by second-order combination, $\bar{r}_n{}^o = 100$, conversion, $\underline{c}$, varies, $\bar{r}_n$ varies with c.

The variation in $\bar{r}_n$ and $\bar{r}_w$ with conversion is shown in Figure 2.3.3 as the solid lines. The weight-average line is a linear function of the conversion. The dispersion ratio, $\bar{r}_w/\bar{r}_n$, increases with conversion as shown in

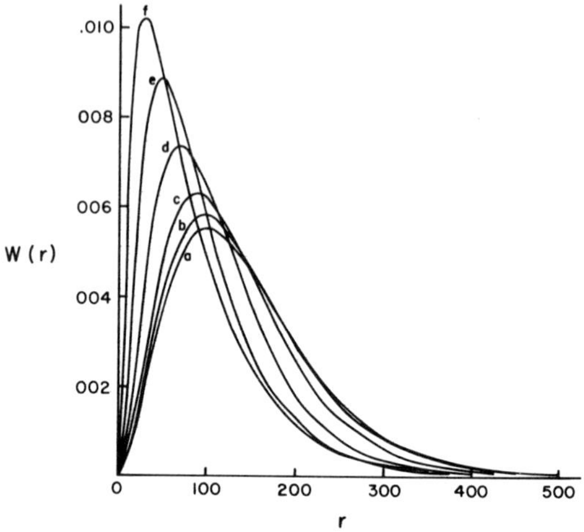

Figure 2.3.2 - Weight distribution as a function of r. Same conditions as in Figure 2.3.1.

Figure 2.3.4. These figures also show the effects of transfer to monomer and to solvent.

If the initial conditions are adjusted to give $\bar{r}_n = 100$ at each conversion, then the distributions of Figures 2.3.5 and 2.3.6 result. There is essentially no change in the distribution until the conversion exceeds 25%.

B. Constant rate of initiation, monomer concentration varies, no transfer, termination by second-order disproportionation.

C. Constant rate of initiation, monomer concentration varies, transfer to solvent only, termination by second-order disproportionation.

Because the disproportionation reaction involves a transfer of an atom or group from one molecule to another in a manner similar to the transfer-to-solvent reaction,

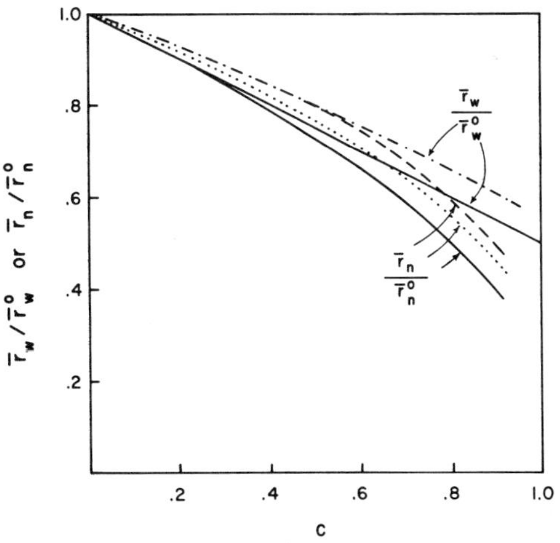

Figure 2.3.3 - The effect of conversion on the number- and weight-average degrees of polymerization, relative to their initial values. Solid lines refer to the following conditions: termination by disproportionation only or with solvent transfer, termination by combination only or with solvent transfer. Dashed line: termination by combination and monomer transfer, $\gamma_m = k_{tr,m}M/\{2R_I\ k_{t,c}\}^{\frac{1}{2}} = 0.25$. Dotted line: termination by disproportionation and monomer transfer $\gamma_m = k_{tr,m}M/\{2R_Ik_{t,d}\}^{\frac{1}{2}} = 0.25$. When $\gamma_m = 0.25$, the dependence of $\bar{r}_w/\bar{r}_w{}^o$ upon conversion is the same for termination by disproportionation or by combination, dash-dotted curve.

both reactions result in the same distribution. We do not include the transfer-to-monomer reaction here because the occurrence of this reaction will vary with conversion. It will be treated separately in Section 4. The equations are:

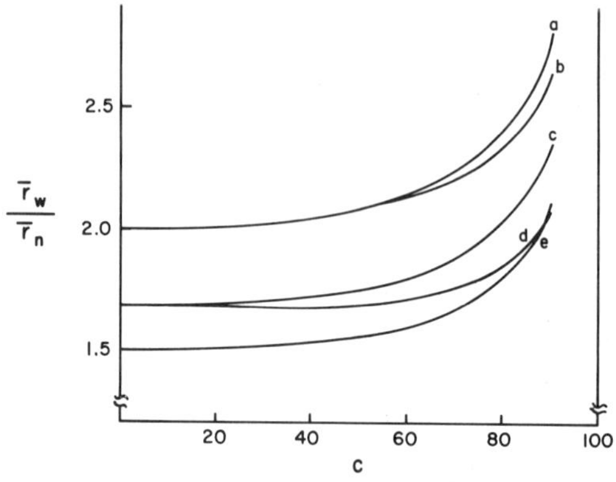

Figure 2.3.4 - The effect of conversion on the dispersion ratio $\bar{r}_w/\bar{r}_n$. a. Termination by disproportionation; transfer to solvent may or may not occur. b. Termination by disproportionation, $\gamma_m = 0.25$. c. Termination by combination $\gamma_s = 0.25$, $\gamma_m = 0$. d. Termination by combination, $\gamma_m = 0.25$, $\gamma_s = 0$. e. Termination by combination, no transfer.

$$W(r) = \frac{r\mu}{(r-1)(M_o - M)}\left\{\left(1 + \frac{\mu}{M_o}\right)^{1-r} - \left(1 + \frac{\mu}{M}\right)^{1-r}\right\} \qquad (5)$$

(Bamford et al., 5) where

$$\mu = [k_{tr,s}S + \{2R_I k_{t,d}\}^{\frac{1}{2}}]/k_p \qquad (6)$$

These distributions are shown in Figures 2.3.7 and 2.3.8 for $\bar{r}_n^{\,o} = 100$ and in Figures 2.3.9 and 2.3.10 for $\bar{r}_n = 100$ for all conversions. Below a conversion of 25%, there is little change in the distributions. The average molecular weights are

Curve	a	b	c	d
c	0.01	0.50	0.75	0.90
$\bar{r}_n^{\,o}$	100.5	139	185	256

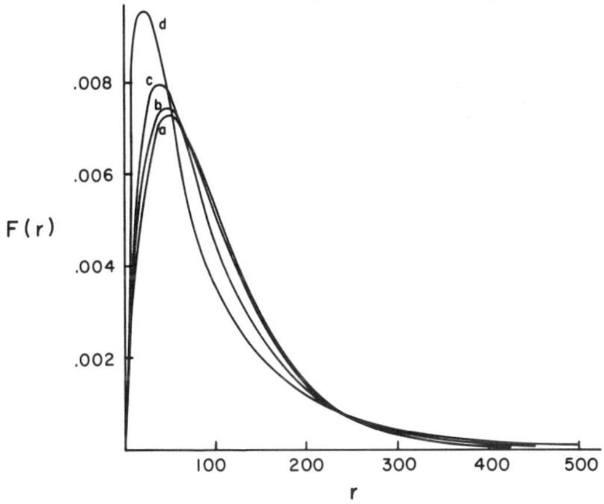

Figure 2.3.5 - Frequency distribution as a function of r. Same conditions as in Figure 2.3.1 except that μ is adjusted to make $\bar{r}_n = 100$ at each conversion. Conversion and $\bar{r}_n^{\,o}$ vary. Compare with Figure 2.3.1. (after Bamford et al., 5).

$$\bar{r}_n = (M_o - M)\, k_p / [\ln (M_o/M)][k_{tr,s}S + \{2R_I k_{t,d}\}^{\frac{1}{2}}] \quad (7)$$

$$\bar{r}_w = (M_o + M)\, k_p / [k_{tr,s}S + \{2R_I k_{t,d}\}^{\frac{1}{2}}] \quad (8)$$

Note, however, that the dependence of $\bar{r}_n$ and $\bar{r}_w$ on conversion is the same as that of Section 3A (compare Figure 2.3.3, solid lines). The variation in the ratio $\bar{r}_w/\bar{r}_n$ with conversion is given in Figure 2.3.4.

Curve	a	b	c	d
c	0.01	0.50	0.75	0.90

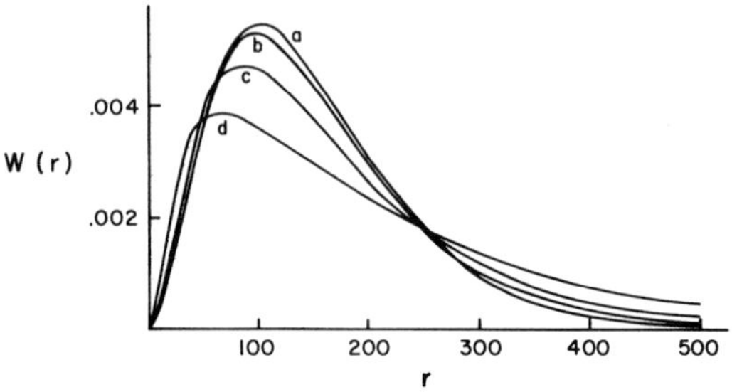

Figure 2.3.6 - Weight distribution as a function of r. Same conditions as Figure 2.3.5. Compare with Figure 2.3.2.

Curve	a	b	c	d	e	f
c	0.01	0.10	0.25	0.50	0.75	0.90
$\bar{r}_n$	99.5	94.8	87.0	72.0	54.1	39.1

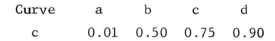

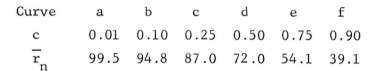

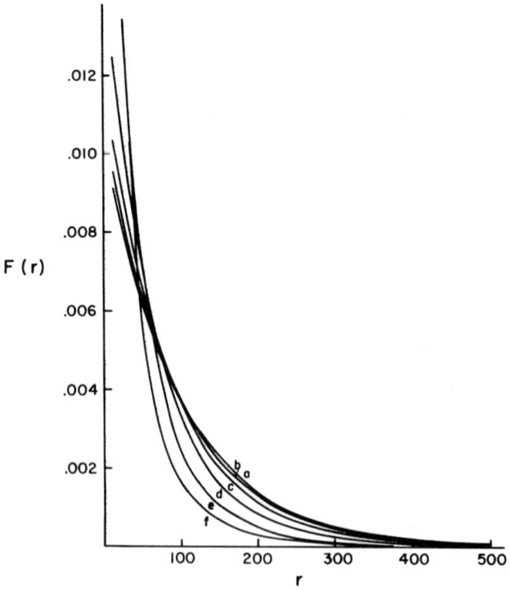

Figure 2.3.7 - Frequency distribution as a function of r

for constant rate of initiation, monomer concentration varies, transfer to solvent only, termination by second order disproportionation, $\bar{r}_n^{\,o} = 100$. Conversion, c, varies, $\bar{r}_n$ varies with c. (after Bamford, et al., 5).

Curve	a	b	c	d	e	f
c	0.01	0.10	0.25	0.50	0.75	0.90

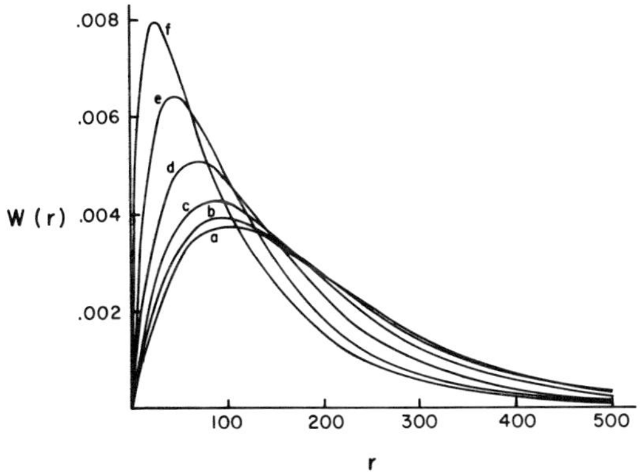

Figure 2.3.8 - Weight distribution as a function of r. Same conditions as in Figure 2.3.7.

D. Constant rate of initiation, monomer concentration varies, transfer to solvent only, termination by second-order combination.

As was done in Section 2D, we define a parameter γ such that

$$\gamma = k_{tr,s}S/\{2R_I k_{t,c}\}^{\frac{1}{2}} \tag{9}$$

The distribution is then

Curve	a	b	c	d
c	0.01	0.50	0.75	0.90
$\bar{r}_n^{\,o}$	100.5	139	185	256

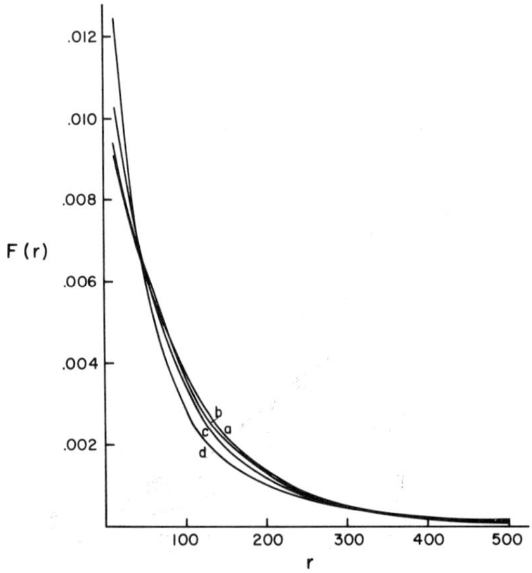

Figure 2.3.9 - Frequency distribution as a function of r. Same conditions as in Figure 2.3.7 except that μ is adjusted to make $\bar{r}_n = 100$ at each conversion. Conversion and $\bar{r}_n^{\,o}$ vary. Compare with Figure 2.3.7. (after Bamford et al., 5).

Curve	a	b	c	d
c	0.01	0.50	0.75	0.90

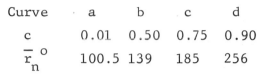

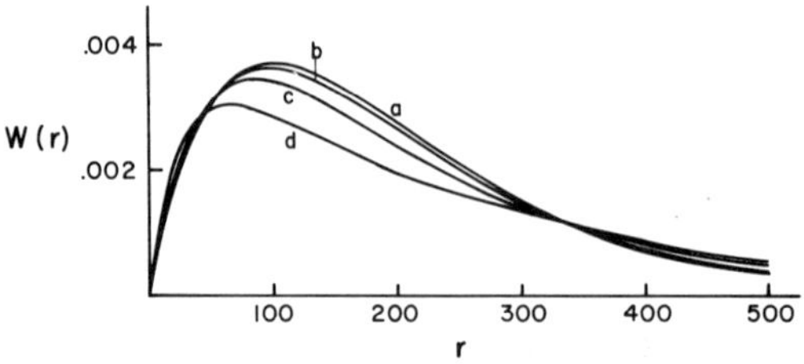

Figure 2.3.10 - Weight distribution as a function of r.

Same conditions as in Figure 2.3.9. Compare with Figure 2.3.8.

$$W(r) = \frac{r\mu}{(1 + \gamma)(M_o - M)} \left\{ \frac{1}{r - 1} \left[1 + \gamma - \frac{(r - 3)}{2(r - 2)} \right] \left[u - v \right] + \frac{(r - 1)\mu}{(r - 2)} \left[\frac{u}{M_o} - \frac{v}{M} \right] \right\} \tag{10}$$

(Bamford et al., 5) where

$$\mu = \{2R_I \, k_{t,c}\}^{\frac{1}{2}} \, (1 + \gamma)/k_p \tag{11}$$

$$u = (1 + \mu/M_o)^{1-r} \tag{12}$$

$$v = (1 + \mu/M)^{1-r} \tag{13}$$

Figures 2.2.1 and 2.2.2 showed the influence of γ on the distribution when $\underline{M}$ is held constant. When $\gamma = 0.25$, the distribution is significantly different from that which occurs when $\gamma = 0$ but still permits second-order termination by recombination to predominate. Therefore, to show the influence of varying monomer concentration upon systems with transfer reactions, we will set γ always equal to 0.25. Intermediate values can be envisioned by comparing the graphs when $\gamma = 0$ and $\underline{M}$ varies and when $\gamma \neq 0$, $\underline{M}$ is constant.

Figures 2.3.11 and 2.3.12 present the distributions for equation (10) for $\gamma = 0.25$, and $\bar{r}_n^{\,o} = 100$. When the initial conditions are selected to give $\bar{r}_n = 100$, the

Curve	a	b	c	d	e	f
c	0.01	0.10	0.25	0.50	0.75	0.90
$\bar{r}_n$	99.5	94.8	87.0	72.0	54.1	39.1

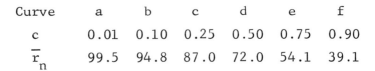

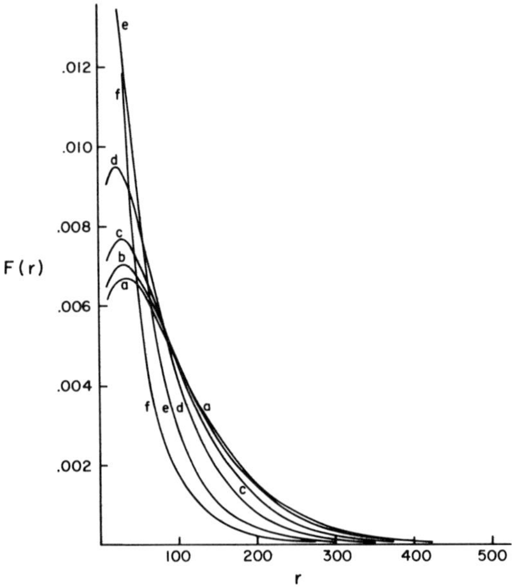

Figure 2.3.11 - Frequency distribution as a function of r for constant rate of initiation, either transfer to solvent only, termination by second-order combination, or no transfer, termination by second-order combination and disproportionation. $\bar{r}_n^{\,o} = 100$, conversion, c, varies. $\gamma = 0.25$. $\bar{r}_n$ varies with c. Compare with Figure 2.3.1. (after Bamford et al., 5).

the distribution is shown in Figures 2.3.13 and 2.3.14. Again, there is little change in the distributions for conversions less than 25%.

The average values of r are

Curve	a	b	c	d	e	f
c	0.01	0.10	0.25	0.50	0.75	0.90

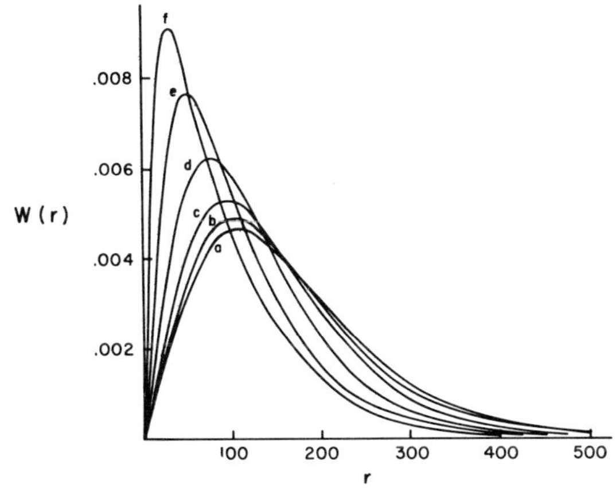

Figure 2.3.12 - Weight distribution as a function of r. Same conditions as in Figure 2.3.11. Compare with Figure 2.3.2.

Curve	a	b	c	d
c	0.01	0.50	0.75	0.90
$\bar{r}_n$	100.5	139	185	256

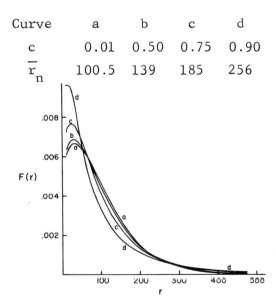

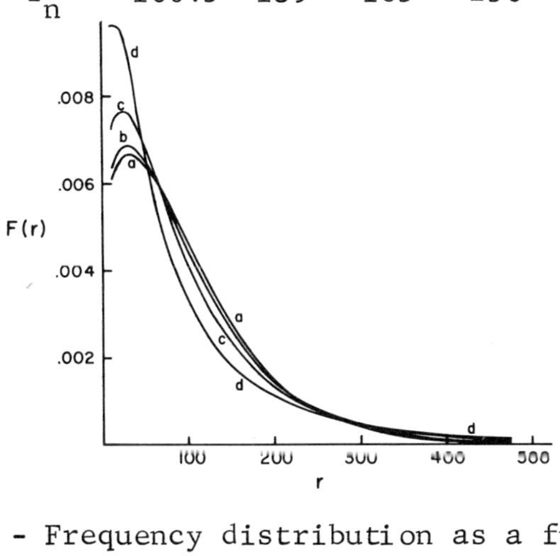

Figure 2.3.13 - Frequency distribution as a function of r.

Same conditions as in Figure 2.3.11 except that μ is adjusted to make $\bar{r}_n = 100$ at each conversion. Conversion and $\bar{r}_n^{\,o}$ vary. The curves for c = 0.01, 0.10, and 0.25 are almost identical.

Curve	a	b	c	d
c	0.01	0.50	0.75	0.90

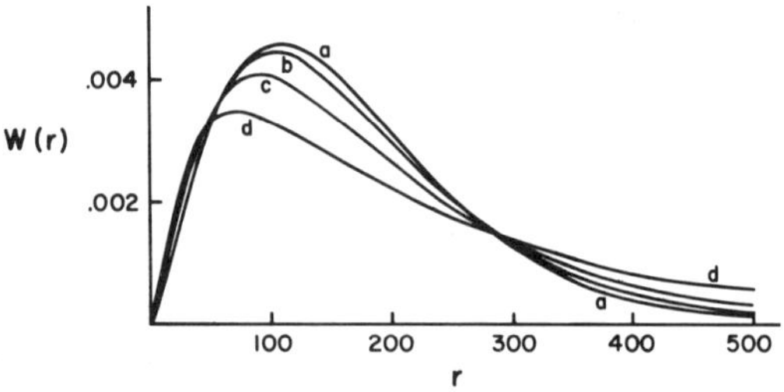

Figure 2.3.14 - Weight distribution as a function of r. Same conditions as in Figure 2.3.13.

$$\bar{r}_n = 2(\gamma + 1)(M_o - M)/(2\gamma + 1)[\ln(M_o/M]\mu \tag{14}$$

$$= 2(M_o - M)\,k_p/(2\gamma + 1)[\ln(M_o/M)]\{2R_I k_{t,c}\}^{\frac{1}{2}}$$

$$\bar{r}_w = (M_o + M)\,k_p(3 + 2\gamma)/2(\gamma + 1)^2\{2R_I k_{r,c}\}^{\frac{1}{2}} \tag{15}$$

Although the distributions are different from those given in Sections 3A and 3C, $\bar{r}_n$, $\bar{r}_w$, and the ratio $\bar{r}_w/\bar{r}_n$ have the identical dependence on conversion, Figures 2.3.3 and 2.3.4.

E. Constant rate of initiation, monomer concentration varies, no transfer, termination by second-order combination and disproportionation (5).

In Section 2E we modified γ to the following definition

$$\gamma = k_{t,d}/k_{t,c} \tag{16}$$

With this definition and by redefining μ as

$$\mu = \{2R_I k_{t,c} \ (1 + \gamma)\}^{\frac{1}{2}}/k_p \tag{17}$$

we can retain the distribution given by equation (10) and shown in Figures 2.3.11 through 2.3.14 for $\gamma = 0.25$. The variables $\underline{u}$ and $\underline{v}$ in equation (10) remain unchanged.

The average values of $\underline{r}$ are

$$\bar{r}_n = 2\ (1 + \gamma)\ (M_o - M)/(2\gamma + 1)\ [\ln(M_o/M)]\mu \tag{18}$$

$$= 2\ (1 + \gamma)^{\frac{1}{2}}(M_o - M)\ k_p/(2\gamma + 1)\ [\ln(M_o/M)]$$

$$\times\ \{2R_I k_{t,c}\}^{\frac{1}{2}}$$

$$\bar{r}_w = (M_o + M)\ k_p(3 + 2\gamma)/2(\gamma + 1)^{3/2}\ \{2R_I k_{t,c}\}^{\frac{1}{2}} \tag{19}$$

The $\bar{r}_n$, $\bar{r}_w$, and the ratio $\bar{r}_w/\bar{r}_n$ are given in Figures 2.3.3 and 2.3.4.

F. Constant rate of initiation, monomer concentration varies, transfer to solvent only, termination by second-order combination and disproportionation.

In Sections 3A to 3E, we separated the transfer-to-solvent reaction from the termination reactions so that the influence of each variable may be clearly seen. The combined distribution function is given by

$$W(r) = \frac{r}{(M_o - M)} \left\{ \frac{\mu}{r-1} - \frac{k_{t,c}(r-3)\{2R_I(k_{t,c} + k_{t,d})\}^{\frac{1}{2}}}{2(k_{t,d} + k_{t,c})(r-1)(r-2)k_p} \right\} (u - v)$$

$$+ \frac{\mu(k_{t,c})(r-1)\{2R_I(k_{t,c} + k_{t,d})\}^{\frac{1}{2}}}{2(k_{t,d} + k_{t,c})(r-2)k_p} \left(\frac{u}{M_o} - \frac{v}{M} \right) \qquad (20)$$

(Bamford et al., 5) where

$$\mu = [k_{tr,s}S + \{2R_I(k_{t,c} + k_{t,d})\}^{\frac{1}{2}}]/k_p \qquad (21)$$

$$u = (1 + \mu/M_o)^{1-r} \qquad (22)$$

$$v = (1 + \mu/M)^{1-r} \qquad (23)$$

$$\bar{r}_n = (M_o - M)k_p/\{[\ln(M_o/M)][k_{tr,s}S$$

$$+ \{R_I/2(k_{t,c} + k_{t,d})\}^{\frac{1}{2}}](k_{t,c} + 2k_{t,d})\} \qquad (24)$$

$$\bar{r}_w = (M_o + M)k_p[k_{tr,s}S$$

$$+ \{R_I/2(k_{t,c} + k_{t,d})\}^{\frac{1}{2}}(3k_{t,c} + 2k_{t,d})]$$

$$\div [k_{tr,s}S + \{2R_I(k_{t,c} + k_{t,d})\}^{\frac{1}{2}}]^2 \qquad (25)$$

4. Monomer Concentration Varies, Transfer-to-Monomer Reaction Occurs

A. Constant rate of initiation, monomer concentration varies, transfer to monomer and to solvent, termination by second-order disproportionation.

Because the transfer-to-monomer reaction depends on the monomer concentration, the preceding sections have neglected this reaction in order to simplify the mathematics. In this section and in the next, we consider the transfer-to-monomer reaction in conjunction with either termination by combination or termination by disproportionation, but not both termination reactions occurring simultaneously. The latter problem must be solved by numerical integration.

The distribution function is

$$W(r) = \frac{\mu}{M_o - M} e^{-\lambda r} \Big[\exp(-\mu r/M_o) - \exp(-\mu r/M)$$

$$+ (r\lambda^2/\mu)[M_o \exp(-\mu r/M_o) - M \exp(-\mu r/M)]$$

$$+ r\lambda(2-\lambda r)\{ Ei(-\mu r/M) - Ei(-\mu r/M_o)\}\Big] \tag{1}$$

(Bamford et al., 5) where

$$\mu = [k_{tr,s}S + \{2R_I\, k_{t,d}\}^{\frac{1}{2}}]/k_p \tag{2}$$

$$\lambda = k_{tr,m}/k_p \tag{3}$$

and

$$\mathrm{Ei}\,(-x) = -\int_x^{\infty} (e^{-u}/u)\, du \tag{4}$$

is the exponential integral. Tabulated values of this integral can be found in mathematical handbooks (8).

To evaluate (1) we define

$$\gamma_m = \frac{k_{tr,m} M_o}{\{2R_I\, k_{t,d}\}^{\frac{1}{2}}} \tag{5}$$

$$\gamma_s = \frac{k_{tr,s} S}{\{2R_I\, k_{t,d}\}^{\frac{1}{2}}} \tag{6}$$

where

$$\lambda = \left[\frac{\gamma_m}{1+\gamma_m + \gamma_s}\right] \frac{1}{\bar{r}_n^{\,o}} \tag{7}$$

and $\bar{r}_n^{\,o}$ is the limiting number-average degree of polymerization as $M \to M_o$

$$\bar{r}_n^{\,o} = \frac{k_p M_o}{k_{tr,m} M_o + k_{tr,s} S + \{2R_I\, k_{t,d}\}^{\frac{1}{2}}} \tag{8}$$

In Figures 2.4.1 and 2.4.2, equation (1) is presented with $\gamma_m = 0.25$, $\gamma_s = 0$, $\bar{r}_n^{\,o} = 100$. These curves differ from those in Figures 2.3.7 and 2.3.8, where $\gamma_s = 0.25$, $\gamma_m = 0$ because the transfer-to-monomer reaction depends on the amount of monomer remaining.

Curve	a	b	c	d	e	f
c	0.01	0.10	0.25	0.50	0.75	0.90
$\bar{r}_n$	99.5	95.8	89.2	76.4	60.0	45.4

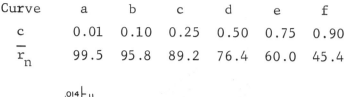

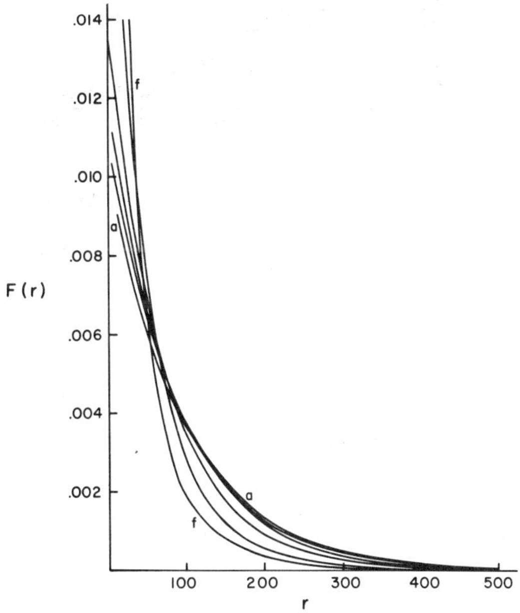

Figure 2.4.1 - Frequency distribution as a function of $\underline{r}$ for constant rate of initiation, monomer concentration varies, transfer to monomer only, termination by second-order disproportionation. $\bar{r}_n^{\,o} = 100$. Conversion, $\underline{c}$, varies. $\gamma_m = 0.25$. $\bar{r}_n$ varies with $\underline{c}$. Compare with Figure 2.3.7. (after Bamford et al., 5).

To find the values of $\bar{r}_n$ and $\bar{r}_w$, we adapt a method used by Tobolsky et al. (9). Assume that within any instant of

Curve	a	b	c	d	e	f
c	0.01	0.10	0.25	0.50	0.75	0.90

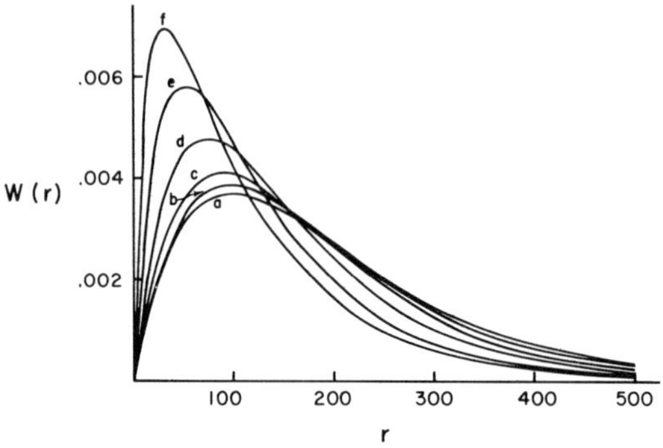

Figure 2.4.2 - Weight distribution as a function of r. Same conditions as in Figure 2.4.1. Compare with Figure 2.3.8.

time, dt, steady state kinetics apply. The number of polymer molecules formed in dt is

$$d\Lambda_0 = R_I + k_{tr,m} MR^* + k_{tr,s} SR^* \tag{9}$$

replacing

R^* by $-dM/k_p M$

and integrating, we have

$$\Lambda_0 = R_I t + \frac{k_{tr,s} S}{k_p} \ln (M_o/M) + \frac{k_{tr,m}}{k_p} (M_o - M) \tag{10}$$

The number of monomers consumed is

$$\Lambda_1 = M_o - M \tag{11}$$

Hence

$$\bar{r}_n = k_p(M_o - M)\Big/\left\{[\{2R_I k_{t,d}\}^{\frac{1}{2}} + k_{tr,s}S]\ \ln(M_o/M) + k_{tr,m}(M_o - M)\right\} \tag{12}$$

At any instant $\underline{dt}$, the distribution of molecules given by $(\bar{r}_w/\bar{r}_n)_i$ is equal to 2, approximately, when termination is by second-order disproportionation. Furthermore, because

$$\bar{r}_w/\bar{r}_n = \frac{\Lambda_2/\Lambda_1}{\Lambda_1/\Lambda_0} \tag{13}$$

We may write for the cumulative distribution at time $\underline{t}$

$$\frac{d\Lambda_2}{dt} = \frac{(\bar{r}_w/\bar{r}_n)_i(d\Lambda_1/dt)^2}{d\Lambda_0/dt} \tag{14}$$

Making the proper substitutions and integrating

$$\bar{r}_w = \Lambda_2/\Lambda_1$$

$$= \frac{2k_p}{k_{tr,m}}\left\{1 - \frac{1 + \gamma_s}{\gamma_m c}\ \ln\left[\frac{1 + \gamma_s + \gamma_m}{1 + \gamma_s + \gamma_m(1 - c)}\right]\right\} \tag{15}$$

where $\underline{c}$ is the conversion.

$$c = (M_o - M)/M_o \tag{16}$$

$\bar{r}_n$ and $\bar{r}_w$ are plotted in Figures 2.3.3 and 2.3.4. The weight-average degree of polymerization $\bar{r}_w$ is not a linear function of conversion when $\gamma_m \neq 0$. The extent of the curvature will depend upon the magnitude of γ_m.

B. Constant rate of initiation, monomer concentration varies, transfer to monomer and to solvent, termination by second-order combination.

The distribution function is

$$W(r) = \frac{re^{-\lambda r}}{M_o - M}\left[\left(\mu - \frac{\{2R_I k_{t,c}\}^{\frac{1}{2}}}{2k_p}\right)\right.$$

$$\times\left\{(e^{-\mu r/M_o} - e^{-\mu r/M})/r\right.$$

$$\left. + \lambda(2 - \lambda r)[Ei(-\mu r/M) - Ei(-\mu r/M_o)\}\right\}$$

$$+ e^{-\mu r/M_o}\left\{\lambda^2 M_o + \frac{\lambda\{2R_I k_{t,c}\}^{\frac{1}{2}}}{k_p} + \frac{\mu\{2R_I k_{t,c}\}^{\frac{1}{2}}}{2k_p M_o}\right\}$$

$$\left. - e^{-\mu r/M}\left\{\lambda^2 M + \frac{\lambda\{2R_I k_{t,c}\}^{\frac{1}{2}}}{k_p} + \frac{\mu\{2R_I k_{t,c}\}^{\frac{1}{2}}}{2k_p M}\right\}\right] \quad (17)$$

(Bamford et al., 5) where

$$\mu = (k_{tr,s}S + \{2R_I k_{t,c}\}^{\frac{1}{2}})/k_p \quad (18)$$

$$\lambda = k_{tr,m}/k_p \quad (19)$$

Ei(-x) is given by equation (4). To evaluate (17),

we define

$$\gamma_m = \frac{k_{tr,m}M_o}{\{2R_I k_{t,c}\}^{\frac{1}{2}}} \qquad \gamma_s = \frac{k_{tr,s}S}{\{2R_I k_{t,c}\}^{\frac{1}{2}}} \tag{20}$$

whence

$$\lambda = \left[\frac{2\gamma_m}{1 + 2(\gamma_m + \gamma_s)}\right]\frac{1}{\bar{r}_n^{\,o}} \tag{21}$$

$$\bar{r}_n^{\,o} = \frac{2k_p M_o}{2k_{tr,m}M_o + 2k_{tr,s}S + \{2R_I k_{t,c}\}^{\frac{1}{2}}} \tag{22}$$

Equation (17) is presented in Figures 2.4.3 and 2.4.4 with $\gamma_m = 0.25$, $\gamma_s = 0$, $\bar{r}_n^{\,o} = 100$. The values of $\bar{r}_n$ and $\bar{r}_w$ are easily calculated by the procedure given in Section 4A, except that $(\bar{r}_w/\bar{r}_n)_i$ is obtained from equations (2.14) and (2.19)

$$(\bar{r}_w/\bar{r}_n)_i = [3\{2R_I k_{t,c}\}^{\frac{1}{2}} + 2k_{tr,s}S + 2k_{tr,m}M]$$

$$\times\ [\{2R_I k_{t,c}\}^{\frac{1}{2}} + 2k_{tr,s}S + 2k_{tr,m}M]$$

$$\div\ 2[\{2R_I k_{t,c}\}^{\frac{1}{2}} + k_{tr,s}S + k_{tr,m}M]^2 \tag{23}$$

which yields

$$\bar{r}_n = 2(M_o - M)k_p/\{[\{2R_I k_{t,c}\}^{\frac{1}{2}}$$

Curve	a	b	c	d	e	f
c	0.01	0.10	0.25	0.50	0.75	0.90
$\bar{r}_n$	99.5	96.4	90.7	79.4	63.8	49.0

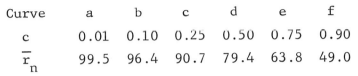

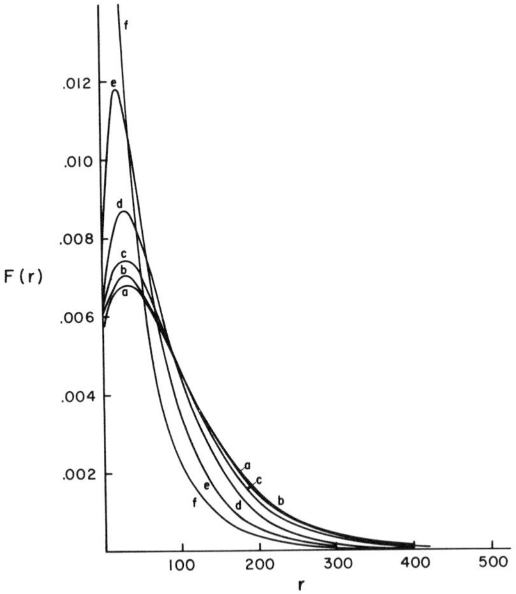

Figure 2.4.3 - Frequency distribution as a function of $\underline{r}$ for constant rate of initiation, monomer concentration varies, transfer to monomer only, termination by second-order combination. $\bar{r}_n^{\,o} = 100$. Conversion, $\underline{c}$, varies. $\gamma_m = 0.25$. $\bar{r}_n$ varies with $\underline{c}$. Compare with Figure 2.3.1. (after Bamford et al., 5).

$$+ 2k_{tr,s}S] \ln (M_o/M) + 2k_{tr,m}(M_o - M)\Big\} \tag{24}$$

$$\bar{r}_w = \frac{(3 + 2\gamma_s)k_pM_o}{\gamma_m^{\,2}\{2R_Ik_{t,c}\}^{\frac{1}{2}}} \left\{\frac{1}{c} \ln \left[\frac{1 + \gamma_s + \gamma_m}{1 + \gamma_s + \gamma_m(1 - c)}\right]\right.$$

Curve	a	b	c	d	e	f
c	0.01	0.10	0.25	0.50	0.75	0.90

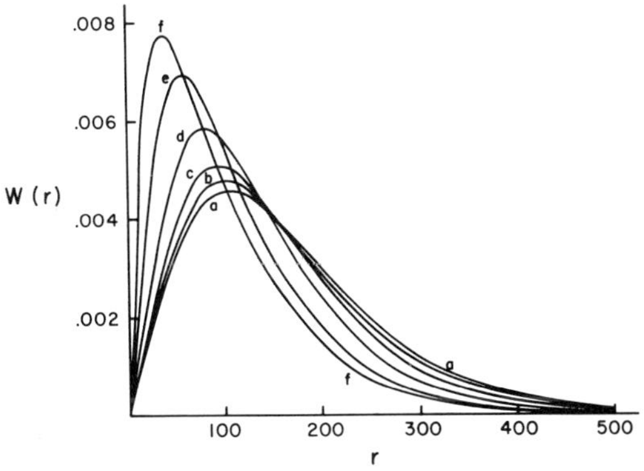

Figure 2.4.4 - Weight distribution as a function of $\underline{r}$. Same conditions as in Figure 2.4.3. Compare with Figure 2.3.2.

$$+ \frac{1}{c}\left[\frac{1 + \gamma_s}{1 + \gamma_s + \gamma_m} - \frac{1 + \gamma_s}{1 + \gamma_s + \gamma_m(1 - c)}\right]\Bigg\}$$

$$+ \frac{2k_pM_o}{\gamma_m\{2R_Ik_{t,c}\}^{\frac{1}{2}}}\left\{1 - \frac{2(1 + \gamma_s)}{\gamma_m c}\ln\left[\frac{1 + \gamma_s + \gamma_m}{1 + \gamma_s + \gamma_m(1 - c)}\right]\right.$$

$$\left. - \frac{1 + \gamma_s}{\gamma_m c}\left[\frac{1 + \gamma_s}{1 + \gamma_s + \gamma_m} - \frac{1 + \gamma_s}{1 + \gamma_s + \gamma_m(1 - c)}\right]\right\} \qquad (25)$$

These equations are shown in Figures 2.3.3 and 2.3.4. The values of γ_s and γ_m selected indicate that $\bar{r}_w/\bar{r}_w^{\,o}$ is almost independent of the mode of termination. However, this is not generally true; as γ_m becomes large, the curves will diverge from each other.

C. Rate of initiation equals $k_i M^2$, monomer concentration varies, transfer to monomer only, termination by second-order combination and disproportionation.

When the rate of initiation equals $k_i M^2$, the differential equation describing dP_n/dt used in the derivation of W(r) is easily integrated to give

$$W(r) = \frac{r(1 - \varsigma)\varsigma^{r-2}}{k_p}\left[k_{tr,m}\varsigma + \left\{\frac{k_i}{2(k_{t,c} + k_{t,d})}\right\}^{\frac{1}{2}} \times \{(1 - \varsigma)(r - 1)k_{t,c} + 2\varsigma k_{t,d}\}\right] \qquad (26)$$

(Bamford et al., 5, Küchler, 4) where

$$\varsigma = k_p/[k_p + k_{tr,m} + \{2k_i(k_{t,c} + k_{t,d})\}^{\frac{1}{2}}] \qquad (27)$$

but this is just equation (2.25) with $R_I = k_i M^2$, $k_{tr,s} = 0$. Note also that W(r) is independent of the conversion. If $k_{t,c} = 0$, the equation reduces to the Schulz-Flory equation (2.6).

5. First-Order Termination or Deactivation

A. Monomer concentration invariant, termination by first-order deactivation.

Let us now consider briefly the effect of first-order termination

$$R_n^* \xrightarrow{k_t} P_n$$

on the molecular weight distribution. As stated in the introduction, the major simplifying assumption is that of steady state kinetics, that is, the rate of initiation of active chains is approximately equal to the rate of destruction of active chains. Thus the total concentration of active chains is invariant with time except only at the start and at the end of the reaction. This assumption imposes a severe restriction on the course of the reaction: a slow deactivation of chains is not permitted because with a constant rate of initiation this would cause the concentration of active chains to increase slowly. Later in this chapter we consider the case where the initiator slowly decreases with time, still maintaining the assumption of steady state kinetics. This restriction is removed in Chapter 3 where "living polymers" are considered. With the steady state assumption, and termination by deactivation, the molecular weight distribution takes the Schulz-Flory form for long chains

$$W(r) = r(1 - \varsigma)^2 \varsigma^{r-1} \tag{1}$$

$$\bar{r}_n = 1/(1 - \varsigma) \tag{2}$$

$$\bar{r}_w = (1 + \varsigma)/(1 - \varsigma) \tag{3}$$

(Küchler, 4) with ς assuming different values depending on the assumed mechanism as shown in Table 2.5.1.

TABLE 2.5.1

Values of ζ for Monomer Concentration Invariant, Termination by First-Order Deactivation

Rate of Initiation	Transfer	Termination	ζ
Constant	Monomer	First-order	$k_pM/(k_pM + k_{tr,m}M + k_t)$
k_iMI	Monomer Solvent	First-order	$k_pM/(k_pM + k_{tr,m}M + k_{tr,s}S + k_t)$
k_iM^2	Monomer	First-order	$k_pM/(k_pM + k_{tr,m}M + k_t)$

Equation 2.5.2 is shown in Figures 1.2.1 and 1.2.2.

B. Monomer concentration varies, termination by first-order deactivation (4,10).

If termination is by a first-order deactivation of chains, and steady-state conditions are assumed, then the resulting distribution is of the Schulz-Flory type when the monomer concentration is allowed to vary

$$W(r) = \frac{\mu}{M_o - M}\left[\left(1 + \frac{\mu}{M_o}\right)^{-r} - \left(1 - \frac{\mu}{M}\right)^{-r}\right] \tag{4}$$

(Jordan and Mathieson, 10) where μ is defined below. Equations of this form were studied in Section 3B, and are shown in Figures 2.3.7 and 2.3.8 (page 76) for $\bar{r}_n^{\,o} = 100$ and in 2.3.9 and 2.3.10 (page 78) for $\bar{r}_n = 100$ for all conversions. In the equation above, we consider

$r >> 1$. The same distribution results if we include the additional termination reaction by solvent

$$R_n^* + S \xrightarrow{k_{tr,s}} P_n \tag{5}$$

The distribution is also independent of the mode of initiation, that is R_I can be constant, $k_i MI$ or $k_i M^2$. The parameter μ can take a number of values depending on the assumed mechanism (Table 2.5.2).

TABLE 2.5.2

Values of μ for Monomer Concentration Varies, Termination by First-Order Deactivation

Rate of Initiation	Transfer	Termination	μ
$k_i MI$	None	First-order	k_t/k_p
$k_i MI$	Solvent	First-order	$(k_t + k_{tr,s}S)/k_p$
$k_i MI$	Solvent	First-order plus solvent termination	$(k_t + k_{tr,s}S + k_{t,s}S)/k_p$
$k_i M^2$	None	First-order	k_t/k_p

$$\bar{r}_n = \frac{M_o - M}{\mu \ln [(M_o + \mu)/(M + \mu)]} \approx \frac{M_o - M}{\mu \ln (M_o/M)} \tag{6}$$

$$\bar{r}_w = 1 + (M_o + M)/\mu \tag{7}$$

when there is no transfer to monomer. Litt (11) gives the average degrees of polymerization when transfer to monomer occurs:

$$\bar{r}_n = \left[\frac{k_{tr,m}}{k_p} + \frac{\mu \ln (M_o/M)}{M_o - M} \right]^{-1} \tag{8}$$

$$\bar{r}_w = (k_p/k_{tr,m}) \times \left\{ 1 - \frac{k_p \mu/k_{tr,m}}{M_o - M} \ln \left[\frac{1 + k_{tr,m} M_o/k_p \mu}{1 + k_{tr,m} M/k_p \mu} \right] \right\} \tag{9}$$

$$\mu = (k_t + k_{tr,s} S)/k_p \tag{10}$$

The distribution equation (4) does not apply when transfer to monomer occurs, as shown in Section 4, because the extent of this reaction depends on the conversion.

C. Rate of initiation equals $k_i MI$, monomer concentration invariant, no transfer, deactivation by the initiator expulsion reaction.

If active chains are deactivated by the reaction

$$R_n^* \xrightarrow{k_{iex}} P_n + I^* \tag{11}$$

where I^* can reinitiate another chain, and stationary state kinetics are assumed, then the Schulz-Flory distribution results, as in Equation (4), but replacing k_t with k_{iex}.

6. Termination by Second-Order Reaction with Monomer (4, 10)

The mechanism of termination is assumed to occur by the reaction

$$R_n^* + M \xrightarrow{k_{t,m}} P_n$$

When stationary-state kinetics are assumed, the effect of conversion disappears; for high molecular weight polymer, the distribution is independent of the conversion. The distribution equation is

$$W(r) = r\lambda^2/(1 + \lambda)^r \tag{1}$$

(Jordan and Mathieson, 10). The average values of r are

$$\bar{r}_n = (1 + \lambda)/\lambda \tag{2}$$

$$\bar{r}_w = (2 + \lambda)/\lambda \tag{3}$$

Again, for high molecular weight polymer, $r - 1 \approx r$, $1/(1 + \lambda) \approx (1 - \lambda)$ so that equation (1) reduces to the Schulz-Flory equation

$$W(r) = r(1 - \zeta)^2\zeta^{r-1}, \qquad \zeta = 1 - \lambda \tag{4}$$

The parameter λ can take several values, depending on the mechanism assumed (Table 2.6.1).

TABLE 2.6.1

Values of λ for Termination by Second-Order Reaction with Monomer

Rate of Initiation	Monomer Concentration Varies	Transfer	λ
Constant	Yes	None	$k_{t,m}/k_p$
k_iMI	Yes	Monomer	$(k_{t,m} + k_{tr,m})/k_p$
Constant	No	Monomer	$(k_{t,m} + k_{tr,m})/k_p$

The distribution function was originally derived by Schulz in 1935 on statistical grounds. In his terminology:

$$W(r)\ dr = r(\ln \zeta)^2\ \zeta^r\ dr \tag{5}$$

(Schulz, 12) where $\zeta = 1 - \lambda$.

7. Two Active Ends per Chain

A. Constant rate of initiation, both ends active, monomer concentration invariant, no transfer, termination by second-order disproportionation.

If the initiator molecule contains two active ends, which are produced by the rate R_I, then the molecular weight distribution is given by

$$W(r) = r^2(1 - \zeta)^3\zeta^{r-1}/(1 + \zeta) \tag{1}$$

(Bamford et al., 5) where

$$\zeta = k_p M / [k_p M + \{2R_I k_{t,d}\}^{\frac{1}{2}}] \quad (2)$$

$$\bar{r}_n = (1 + \zeta)/(1 - \zeta) \quad (3)$$

$$\bar{r}_w = (1 + 4\zeta + \zeta^2)/(1 - \zeta)(1 + \zeta) \quad (4)$$

The mole fraction distribution F(r) is just the Schulz-Flory weight distribution for molecules with only one active end,

$$F(r) = r(1 - \zeta)^2 \zeta^{r-1}$$

The mole fraction distribution for this case is given in Figure 1.2.2, with the $\bar{r}_n$ values given in that Figure are just one-half the $\bar{r}_n$ values for this distribution. Thus the curve W(r), $\bar{r}_n = 50$ in Figure 1.2.2 corresponds exactly with F(r), $\bar{r}_n = 100$ for the present distribution. The weight fraction distribution for (1) is given in Figure 2.7.1.

Bamford and Tompa (13) give the distribution equation when the transfer-to-monomer reaction occurs.

B. Constant rate of initiation, both ends active, monomer concentration invariant, transfer to monomer only, termination by second-order combination.

If the initiator molecule contains two active ends which are produced by the rate R_I, and termination is by combination, the distribution is of the Schulz-Flory form

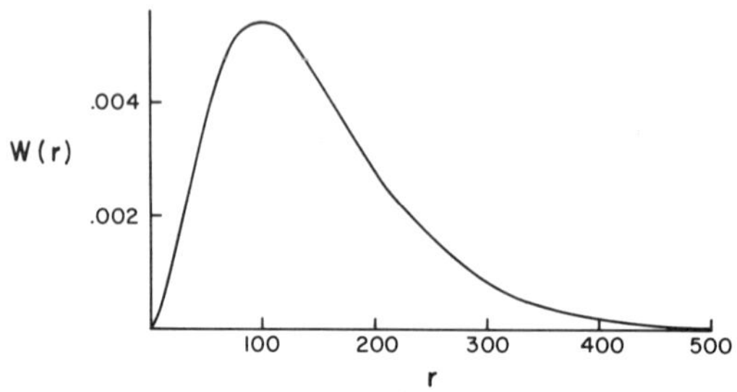

Figure 2.7.1 - Weight distribution as a function of r for constant rate of initiation, both ends active, no transfer, monomer concentration invariant, termination by second-order disproportionation, $\bar{r}_n = 100$. The frequency distribution is given in Figure 1.2.2 with $\bar{r}_n = 50$. (after Bamford et al., 5).

$$W(r) = (r/\bar{r}_n)^2 \exp(-r/\bar{r}_n) \tag{5}$$

(Bamford et al., 5) where

$$\bar{r}_n - (k_pM + k_{tr,m}M)/k_{tr,m}M = 1/(1 - \zeta) \tag{6}$$

$$\bar{r}_w = (1 + \zeta)\bar{r}_n \tag{7}$$

Note that the distribution equation does not contain $k_{t,c}$ because combination of two chains both containing two active ends results in one chain with both ends active. The only means of forming inactive chains is by combination of two monoactive chains. Such chains are only formed by the transfer-to-monomer reaction.

8. Slow Exhaustion of Initiator, Nonsteady State Kinetics

A. Rate of initiation equals $k_i I$, monomer concentration invariant, no transfer, termination either by second-order disproportionation or by second-order combination.

Chien (3, 14) has considered the case when the initiator is slowly exhausted as polymerization continues. The total concentration of active centers is given by

$$R^* = x[k_i I_o/2k_{t,d}]^{\frac{1}{2}} \left[\frac{\tilde{K}_1(ax) - A\tilde{I}_1(ax)}{\tilde{K}_0(ax) + A\tilde{I}_0(ax)}\right] \tag{1}$$

where

$$x = \exp\ (-k_i t/2) \tag{2}$$

$$A = \tilde{K}_1(a)/\tilde{I}_1(a) \tag{3}$$

$$a = 2(2k_{t,d} I_o/k_i)^{\frac{1}{2}} \tag{4}$$

The $\tilde{K}_0$, $\tilde{K}_1$, $\tilde{I}_0$, and $\tilde{I}_1$ are the modified Bessel functions of zero and first order respectively. The number- and weight-average degrees of polymerization, <u>derived without steady state assumptions</u>, are

$$\bar{r}_n = 1 + \frac{k_p M_o}{2k_{t,d} I_o(1 - x^2)} \ln\left[\frac{\tilde{K}_0(ax) + A\tilde{I}_0(ax)}{\tilde{K}_0(a) + A\tilde{I}_0(a)}\right] \tag{5}$$

$$\bar{r}_w = \left\{ I_o(1 - x^2) + \frac{3k_pM_o}{2k_{t,d}} \ln\left[\frac{\tilde{K}_0(ax) + A\tilde{I}_0(ax)}{\tilde{K}_0(a) + A\tilde{I}_0(a)}\right]\right.$$

$$- \frac{2k_p{}^2M_o{}^2}{k_ik_{t,d}}\left(\ln x\right.$$

$$\left.\left. - [\tilde{K}_0(a) + \tilde{I}_0(a)] \int_1^x \frac{dw}{w[\tilde{K}_0(ax) + A\tilde{I}_0(ax)]}\right)\right\}$$

$$\div \left\{ I_o(1 - x^2) + \frac{k_pM_o}{2k_{t,d}} \ln\left[\frac{\tilde{K}_0(ax) + A\tilde{I}_0(ax)}{\tilde{K}_0(a) + A\tilde{I}_0(a)}\right]\right\} \tag{6}$$

to obtain R* as a function of time when termination is by combination replace $k_{t,d}$ in equation (1) with $k_{t,c}$.

$$\bar{r}_n = \frac{k_pM_o}{k_{t,c}[I_o(1 - x^2) + R^*]} \ln\left[\frac{\tilde{K}_0(ax) + A\tilde{I}_0(ax)}{\tilde{K}_0(a) + A\tilde{I}_0(a)}\right] \tag{7}$$

$\bar{r}_w$ is not given.

Because of the complexity of these equations, no graphs are presented for $\bar{r}_n$ and $\bar{r}_w$ as a function of time or initiator concentration. Bamford has derived simpler equations which are considered in the next Section.

9. Dead-End Polymerization, Monomer Concentration Invariant

"Dead-end polymerization" is the process described by Tobolsky (15) wherein the initiator becomes exhausted before the monomer becomes exhausted and polymerization ceases at an intermediate conversion. From a consideration of the equation,

$$(I_o^{\frac{1}{2}} - I^{\frac{1}{2}}) = \left(\frac{k_{t,d}k_d}{f}\right)^{\frac{1}{2}} \frac{\ln(M_o/M)}{2k_p} \qquad (1)$$

which comes from (dI/dt)/(dM/dt), complete conversion cannot occur, regardless of the values of the assorted constants. Here, f is the efficiency of initiation. Essentially complete conversion may occur, at say 99%, with very little initiator decomposition if the proper polymerization conditions are selected. These situations have already been examined. The distribution functions have not been obtained when both the monomer and initiator concentrations are varying, so we shall first examine the effect of varying the initiator concentration while maintaining the monomer concentration invariant. This situation is not dead-end polymerization as described by Tobolsky, but it is instructive to see how the distributions and the molecular weight averages vary with initiator concentration. The latter can be directly compared to the molecular weight averages obtained when the monomer concentration varies (Section 10).

A. No transfer, termination by disproportionation.

Bamford (2) considered the case of rapidly decaying initiators

$$I \rightarrow nR^* \qquad (2)$$

$$I = I_o \exp(-k_d t) \tag{3}$$

under the conditions of constant monomer concentration and that within a small time interval, dt, the steady-state assumption is valid. The weight distribution function at a final concentration of initiator, I, is

$$W(r) = \frac{\bar{r}_n^{\,o}}{r^2(1 - y_o)} \left\{ (x^2 y_o^{\,2} + 2xy_o + 2) \exp(-xy_o) - (x^2 + 2x + 2) \exp(-x) \right\} \tag{4}$$

(Bamford, 2) where

$$x = r/\bar{r}_n^{\,o} \tag{5}$$

$$y_o = (I/I_o)^{\frac{1}{2}} \tag{6}$$

$$\bar{r}_n^{\,o} = k_p M/(2nk_d k_{t,d} I_o)^{\frac{1}{2}} \tag{7}$$

$$= \bar{r}_n \text{ with } y_o = 1$$

and n is the number of initiating species which result from the decomposition of I.

$$\bar{r}_n = 2\, k_p M/(2k_{t,d} nk_d I_o)^{\frac{1}{2}} (1 + y_o) \tag{8}$$

$$\bar{r}_w = -2k_p M \ln y_o/(2nk_{t,d} k_d I_o)^{\frac{1}{2}} (1 - y_o) \tag{9}$$

Note that $\bar{r}_w$ approaches infinity as $(I/I_o)^{\frac{1}{2}} = y_o$ approaches zero. The distribution function is shown in Figure 2.9.1 for several values of I/I_o. Values of $\bar{r}_n$ and $\bar{r}_w$ are given in Figure 2.9.2.

Curve	a	b	c	d	e	f	g
I/I_o	1.00	0.75	0.50	0.25	0.10	0.01	0

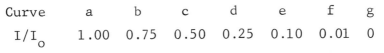

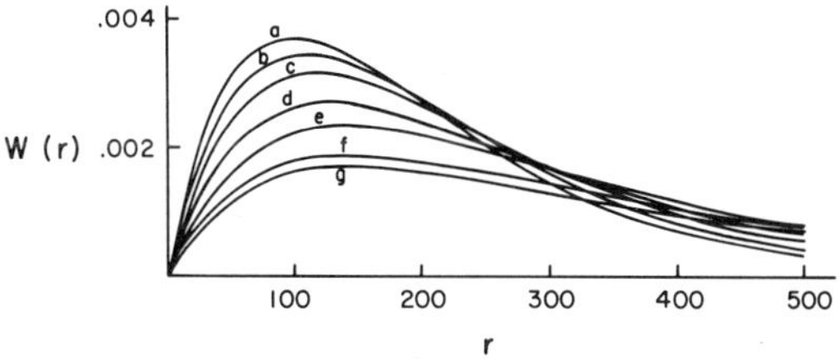

Figure 2.9.1 - Weight distribution as a function of r for dead-end polymerization, no transfer, monomer concentration invariant, termination by second-order disproportionation, I/I_o varies, $\bar{r}_n^{\,o} = 100$. (after Bamford, 2).

B. Transfer to monomer and to solvent, termination by disproportionation.

The effect of the transfer reactions upon the distribution in Equation (4) can be described by replacing y_o and $\bar{r}_n^{\,o}$, equations (6) and (7) by

$$y = [\gamma + (I/I_o)^{\frac{1}{2}}]/(1 + \gamma) \tag{10}$$

where

$$\gamma = \frac{k_{tr,m}M + k_{tr,s}S}{\{2k_{t,d}nk_dI_o\}^{\frac{1}{2}}} \tag{11}$$

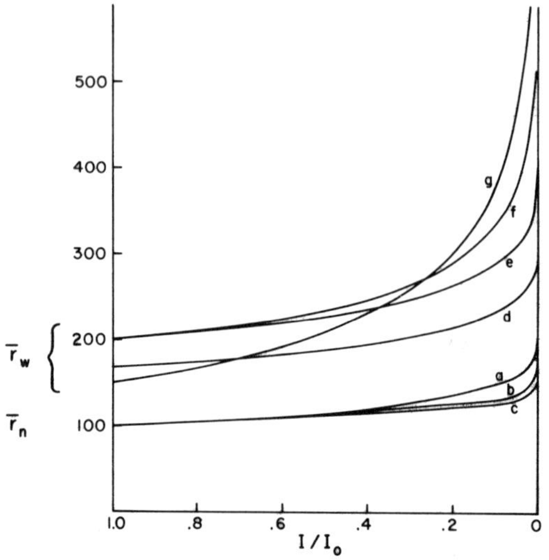

Figure 2.9.2 - Number- and weight-average degrees of polymerization as a function of I/I_o, monomer concentration invariant. $\bar{r}_n$: (a) termination by disproportionation or combination, no transfer, (b) termination by disproportionation with transfer, $\gamma = 0.25$, (c) termination by combination with transfer, $\gamma = 0.25$; $\bar{r}_w$: (d) termination by combination, no transfer, (e) termination by combination with transfer, $\gamma = 0.25$, (f) termination by disproportionation, no transfer, (g) termination by disproportionation with transfer, $\gamma = 0.25$.

and

$$\bar{r}_n = 2\, k_p M/\{2k_{t,d} n k_d I_o\}^{\frac{1}{2}}\, [2\gamma + 1 + y_o] \tag{12}$$

$$\bar{r}_n{}^o = \bar{r}_n \text{ with } y_o = 1 \tag{13}$$

$$\bar{r}_w = -2k_p M[\ln y]/(2k_{t,d}\, n k_d I_o)^{\frac{1}{2}}(1 - y) \tag{14}$$

The effect of transfer on the molecular weights is given in Figure 2.9.2. In this case $\bar{r}_w$ does not approach infinity as I/I_o approaches zero because the value of y remains finite.

C. No transfer, termination by combination.

The distribution function is

$$W(r) = \frac{\bar{r}_n^{\,o}}{4r^2(1 - y_o)} \{[x^3 y_o^{\,3} + 3x^2 y_o^{\,2} + 6xy_o + 6]$$

$$\times \exp(-xy_o) - [x^3 + 3x^2 + 6x + 6] \exp(-x)\} \qquad (15)$$

(Bamford, 2) where

$$x = r/\bar{r}_n \qquad (16)$$

$$\bar{r}_n^{\,o} = \bar{r}_n \text{ with } y_o = 1 \qquad (17)$$

$$y_o = (I/I_o)^{\frac{1}{2}} \qquad (18)$$

$$\bar{r}_n = 4k_p M/(2k_{t,c} n k_d I_o)^{\frac{1}{2}}(1 + y_o) \qquad (19)$$

$$\bar{r}_w = -3k_p M(\ln y_o)/[2k_{t,c}\, n k_d I_o]^{\frac{1}{2}}\,(1 - y_o) \qquad (20)$$

The distribution is given in Figure 2.9.3, while the molecular weights are given in Figure 2.9.2.

D. Transfer to monomer and to solvent, termination by combination.

Curve	a	b	c	d	e	f	g
I/I_o	1.00	0.75	0.50	0.25	0.10	0.01	0

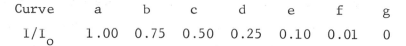

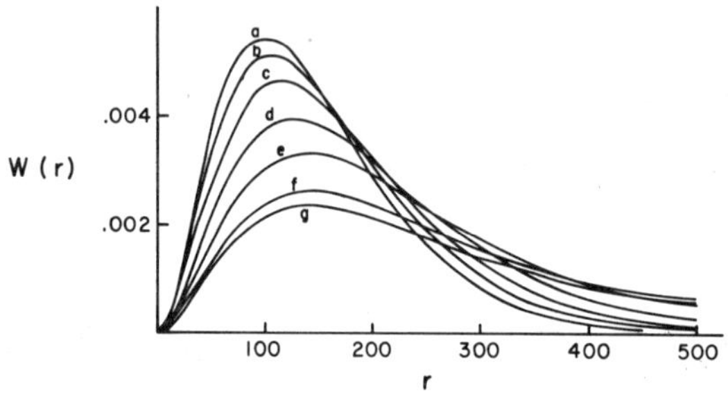

Figure 2.9.3 - Weight distribution as a function of r for dead-end polymerization, no transfer, monomer concentration invariant, termination by second-order combination, I/I_o varies, $\bar{r}_n^{\,o} = 100$.

To obtain the distribution function for transfer reactions with termination by combination, first define the following functions:

$$y = [\gamma + y_o]/(2\gamma + 1) \tag{21}$$

$$y_1 = (\gamma + 1)/(2\gamma + 1) \tag{22}$$

$$y_o = (I/I_o)^{\frac{1}{2}} \tag{23}$$

where

$$\gamma = (k_{tr,m}M + k_{tr,s}S)/\{2k_{t,c}nk_dI_o\}^{\frac{1}{2}} \tag{24}$$

$$x = 2r/\bar{r}_n \tag{25}$$

then

$$W(r) = \frac{\bar{r}_n^{\,o}}{4r^2(y_1 - y)} \{[x^3y^3 + 3x^2y^2 + 6xy + 6]\exp(-xy)$$

$$- [x^3y_1^{\,3} + 3x^2y_1^{\,2} + 6xy_1 + 6]\exp(-xy_1)$$

$$- x^3[\gamma/(2\gamma + 1)][y^2\exp(-xy) - y_1^{\,2}\exp(-xy_1)]\} \qquad (26)$$

(Bamford, 2)

$$\bar{r}_n = 4k_pM/\{2k_{t,c}\,nk_dI_o\}^{\frac{1}{2}}[4\gamma + 1 + y_o] \qquad (27)$$

$$\bar{r}_n^{\,o} = \bar{r}_n \text{ with } (I/I_o)^{\frac{1}{2}} = y_o = 1 \qquad (28)$$

$$\bar{r}_w = \frac{3k_pM}{(2k_{t,c}nk_dI_o)^{\frac{1}{2}}(1 - y_o)} \left\{\ln(y_1/y)\right.$$

$$\left. - \frac{\gamma}{3(2\gamma + 1)}\left[\frac{1}{y} - \frac{1}{y_1}\right]\right\} \qquad (29)$$

The distribution is given in Figure 2.9.4 and the molecular weights in Figure 2.9.2. When transfer reactions occur, $\bar{r}_w$ remains finite as y_o approaches zero.

E. Active radicals produced by second-order decay of initiator, no transfer, termination by disproportionation.

If the reaction of initiator molecules to form n active centers is

$$2I_o \rightarrow nR^* \qquad (30)$$

Curve	a	b	c	d	e	f	g
I/I_o	1.00	0.75	0.50	0.25	0.10	0.01	0

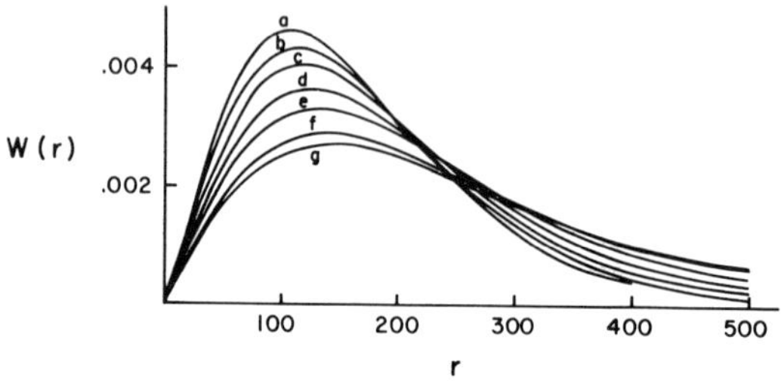

Figure 2.9.4 - Weight distribution as a function of $\underline{r}$ for dead-end polymerization, monomer concentration invariant, transfer to monomer and to solvent, termination by second-order combination, I/I_o varies, $\bar{r}_n^{\,o} = 100$, $\gamma = 0.25$. Compare with Figure 2.9.3.

then the initiator concentration varies with time as

$$1/I - 1/I_o = k_2 t \tag{31}$$

For termination by disproportionation

$$W(r) = \{(1 + xy)\exp(-xy) - (1 + x)\exp(-x)\} \div [r \ln(1/y)] \tag{32}$$

(Bamford, 2) where

$$x = r/\bar{r}_n^{\,o} \qquad y = I/I_o \tag{33}$$

$$\bar{r}_n^{\,o} = k_p M/\{2k_{t,d}\, n k_2 I_o^{\,2}\}^{\frac{1}{2}} \tag{34}$$

$$\bar{r}_n = \bar{r}_n^{\,o}\,[\ln(I_o/I)]/[1 - (I/I_o)] \tag{35}$$

$$\bar{r}_w = 2\bar{r}_n^{\,o}\,[(I_o/I) - 1]/[\ln(I_o/I)] \tag{36}$$

The distribution function is given in Figure 2.9.5. Both $\bar{r}_n$ and $\bar{r}_w$ approach infinity as I approaches zero, Figure 2.9.6.

Curve	a	b	c	d
I/I_o	1.00	0.50	0.10	0.001

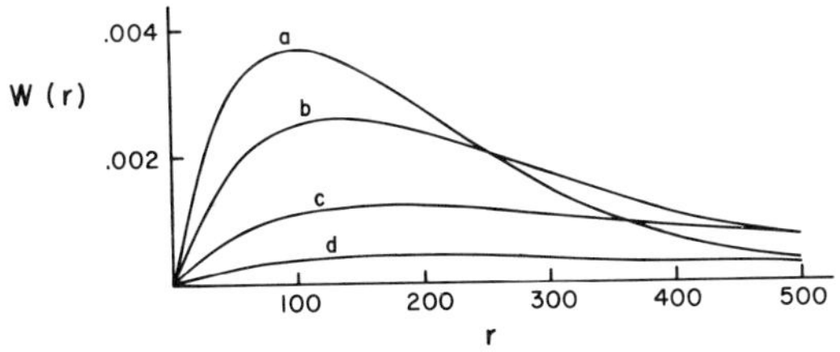

Figure 2.9.5 - Weight distribution as a function of r for dead-end polymerization, monomer concentration invariant, no transfer, active molecules produced by second-order decay of initiator $2I \rightarrow nR^*$, termination by second-order disproportionation, I/I_o varies, $\bar{r}_n^{\,o} = 100$, $\bar{r}_n \rightarrow \infty$ as $I \rightarrow 0$. (after Bamford, 2).

F. Active radicals produced by second-order decay of initiator, no transfer, termination by combination.

The distribution equation is

$$W(r) = \{(x^2y^2 + 2xy + 2)\exp(-xy)$$

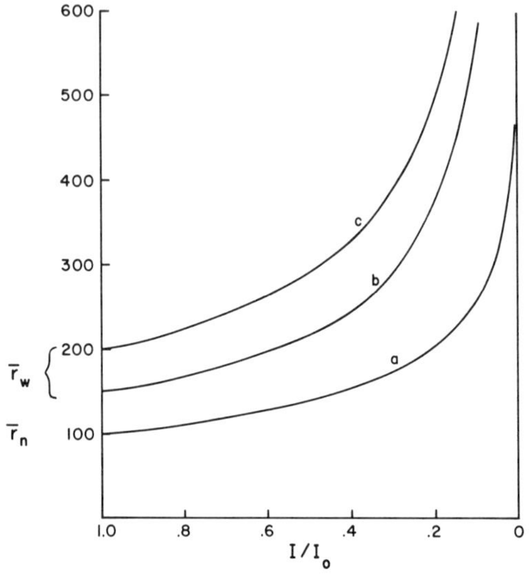

Figure 2.9.6 - Number- and weight-average degrees of polymerization as a function of I/I_o, monomer concentration invariant. $\bar{r}_n$: (a) termination by disproportionation or by combination, no transfer; $\bar{r}_w$: (b) termination by combination, no transfer, (c) termination by disproportionation, no transfer.

$$- (x^2 + 2x + 2) \exp(-x)\}/[2r\ln(1/y)] \tag{37}$$

(Bamford, 2) where

$$x = 2r/\bar{r}_n^{\,o} \tag{38}$$

$$y = I/I_o \tag{39}$$

$$\bar{r}_n^{\,o} = 2k_p M / \{2k_{t,c}\, nk_2 I_o^{\,2}\}^{\frac{1}{2}} \qquad (40)$$

$$\bar{r}_n = \bar{r}_n^{\,o}\, [\ln(I_o/I)]/[1 - (I/I_o)] \qquad (41)$$

$$\bar{r}_w = 3\, \bar{r}_n^{\,o}\, [(I_o/I) - 1]/[\ln(I_o/I)] \qquad (42)$$

The distribution function is given in Figure 2.9.7. Again both $\bar{r}_n$ and $\bar{r}_w$ approach infinity as $\underline{I}$ approaches zero (Figure 2.9.6).

Curve	a	b	c	d
I/I_o	1.00	0.50	0.10	0.001

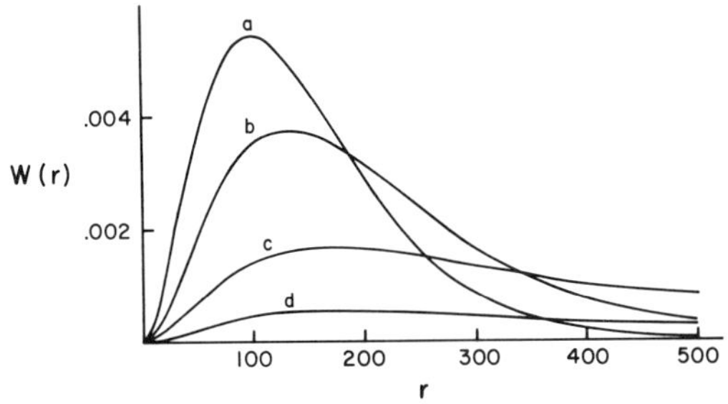

Figure 2.9.7 - Weight distribution as a function of $\underline{r}$. Same conditions as in Figure 2.9.5 except termination by second-order combination.

10. Dead-End Polymerization, Monomer Concentration Varies

A. Transfer to monomer and to solvent, termination by disproportionation.

The average molecular weights for dead-end polymerization can be calculated by use of the procedure outlined in Section 2.4. This is essentially the procedure given in reference 9, except that an error exists in the derivation given therein.* For an initiator which yields *2f* active radicals by first order decay,

$$\bar{r}_n = \frac{M_o c}{2fI_o(1 - y^2) - C_s S \ln(1 - c) + C_m M_o c} \qquad (1)$$

$$\bar{r}_w = \frac{2M_o}{c} \int_0^c \frac{(1 - c)\, dc}{2fKI_o^{\frac{1}{2}} y + C_s S + C_m M_o (1 - c)} \qquad (2)$$

(Tobolsky, 9) where

$$y = 1 + \frac{K}{2I_o^{\frac{1}{2}}} \ln(1 - c) = (I/I_o)^{\frac{1}{2}} \qquad (3)$$

$$C_s = k_{tr,s}/k_p \qquad C_m = k_{tr,m}/k_p \qquad (4)$$

$$c = (M_o - M)/M_o \qquad K = (k_{t,d} k_d / f)^{\frac{1}{2}}/k_p \qquad (5)$$

---

* Equation 7 of reference 9 gives the weight fraction of polymer formed in the interval *dt* as $-dM/(M_o - M)$ where *M* in the denominator is treated as a variable. The correct weight fraction is $-dM/(M_o - M_t)$ where M_t is the monomer concentration at time *t* and is a constant.

f = efficiency of initiation S = solvent (6)

The effect of exhausting the initiator at intermediate conversions on the molecular weights is shown in Figure 2.10.1 for termination by disproportionation with no

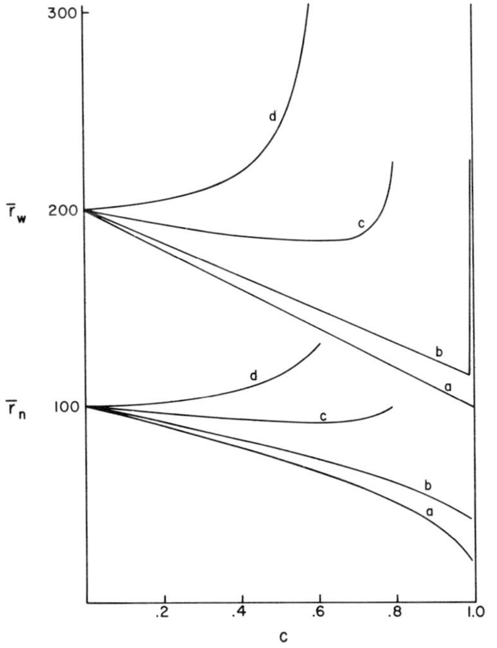

Figure 2.10.1 - Number- and weight-average degrees of polymerization as a function of conversion for dead-end polymerization with monomer concentration varying, no transfer, and termination by disproportionation. To show only the effect of various final conversions on the molecular weights with a constant initial value, $\bar{r}_n^{\,o} = 100$, an artificial adjustment had to be made in the parameter $K = (k_{t,d}k_d/f)^{\frac{1}{2}}/2k_pI_o^{\frac{1}{2}}$. (a) constant rate of initiation

(not dead-end polymerization), (b) $c_f = 0.99$, (c) $c_f = 0.80$, (d) $c_f = 0.60$. (after Tobolsky, 9).

transfer. This is an artificial situation because to require $\bar{r}_n^{\,o}$ to be 100 while making adjustments in c_f, the conversion at infinite time,

$$c_f = 1 - \exp[- 2I_o^{\frac{1}{2}}k_p/\{k_{t,d}k_d/f\}^{\frac{1}{2}}] \tag{7}$$

requires adjustments in the temperature-dependent parameters k_p, $k_{t,d}$, k_d, and $\underline{f}$. The graph does show that dead-end polymerization will have quite drastic effects on the molecular weight distribution. At high values of c_f, the molecular weights initially decrease with increasing conversion increasing only as c_f is approached, while at low values of c_f the molecular weights increase with increasing conversion.

B. Transfer to monomer and to solvent, termination by combination.

$$\bar{r}_n = \frac{M_oc}{fI_o(1 - y^2) - C_sS \ln (1 - c) + C_mM_oc} \tag{8}$$

$$\bar{r}_w = \frac{M_o}{c}\int_0^c \frac{[6fKI_o^{\frac{1}{2}}y + 2C_sS + 2C_mM_o(1 - c)](1 - c)\, dc}{[2fKI_o^{\frac{1}{2}}y + C_sS + C_mM_o(1 - c)]^2} \tag{9}$$

where

$\underline{y}$, C_s, C_m, $\underline{c}$, and $\underline{f}$ are the same as in Section 10A.

$$K = (k_{t,c} k_d / f)^{\frac{1}{2}} / k_p$$

These equations are plotted in Figures 2.10.2 against conversion for $\gamma_s = \gamma_m = 0$; $\gamma_s = 0$, $\gamma_m = 0.25$; and $\gamma_s = 0.25$, $\gamma_m = 0$.
where

$$\gamma_s = \frac{k_{tr,s} S}{\{4 f k_d k_{t,c} I_o\}^{\frac{1}{2}}} \qquad \gamma_m = \frac{k_{tr,m} M_o}{\{4 f k_d k_{t,c} I_o\}^{\frac{1}{2}}}$$

or the similar parameters replacing $k_{t,c}$ by $k_{t,d}$ for the initial condition of $\bar{r}_n^{\,o} = 100$ and further that the initiator is expended at 60% conversion. The equations are also plotted against I/I_o in Figure 2.10.3 for comparison with Figure 2.9.2.

11. Copolymerization

A. Constant rate of initiation, monomer concentration invariant, transfer to monomer and to solvent, termination by second-order disproportionation, two different monomers present.

When two monomers, M_1 and M_2 are present, there are four propagation constants:

$$\sim\sim\sim M_1^* + M_1 \rightarrow \sim\sim\sim M_1^* \qquad k_{11}$$

$$\sim\sim\sim M_1^* + M_2 \rightarrow \sim\sim\sim M_2^* \qquad k_{12}$$

Curve	Transfer	Termination
a	None	Disproportionation and Combination, Not Dead-End
b	Solvent	Combination
c	Solvent	Disproportionation
d	Monomer	Combination
e	Monomer	Disproportionation
f	None	Combination
g	None	Disproportionation

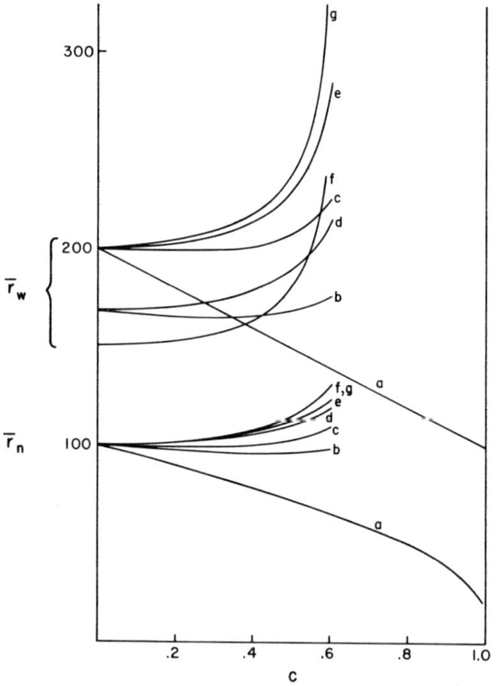

Figure 2.10.2 - Number- and weight-average degrees of polymerization as a function of conversion for dead-end polymerization with varying monomer concentration and $c_f = 0.60$. (after Tobolsky, 9).

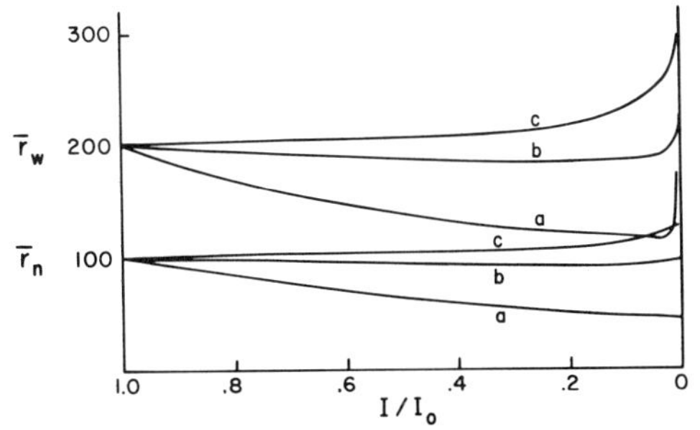

Figure 2.10.3 - Number- and weight-average degrees of polymerization as a function of the remaining initiator fraction with varying monomer concentration, varying final monomer concentration, no transfer, termination by disproportionation. (a) $c_f = 1.00$, (b) $c_f = 0.80$, (c) $c_f = 0.60$. (after Tobolsky, 9).

$$\begin{array}{ll} \sim\sim M_2^* + M_1 \rightarrow \sim\sim M_1^* & k_{21} \\ \sim\sim M_2^* + M_2 \rightarrow \sim\sim M_2^* & k_{22} \end{array} \tag{1}$$

where k_{ij} is the propagation constant for addition of monomer j to radical i to make a new radical one monomer unit larger of type j.

With the assumption of steady state kinetics and that

$$k_{12}M_1^*M_2 = k_{21}M_2^*M_1 \tag{2}$$

the instantaneous composition of polymer is given by

$$\frac{dM_1}{dM_2} = \frac{M_1(r_1M_1 + M_2)}{M_2(M_1 + r_2M_2)} \tag{3}$$

where $r_1 = k_{11}/k_{12}$, $r_2 = k_{22}/k_{21}$, are the "reactivity ratios." Equation (3) has been discussed numerous times and need not be repeated here. Tables of r_1 and r_2 may be found in Ham (16) and in Brandrup and Immergut (17). Discussions of the integrated form of equation (3) may be found in Küchler (4) and in Alfrey, Bohrer, and Mark (18). Our main interest is in the compositional distributions of the chains as they are formed. Just as there are four rate constants of propagation, so also are there numerous constants for termination by disproportionation:

$$\begin{aligned}
&\sim\!\sim\!\sim M_1^* + \sim\!\sim\!\sim M_1^* \rightarrow \sim\!\sim\!\sim M_1 + \sim\!\sim\!\sim M_1 (DB)\\
&\sim\!\sim\!\sim M_1^* + \sim\!\sim\!\sim M_2^* \rightarrow \sim\!\sim\!\sim M_1 + \sim\!\sim\!\sim M_2 (DB)\\
&\sim\!\sim\!\sim M_1^* + \sim\!\sim\!\sim M_2^* \rightarrow \sim\!\sim\!\sim M_1 (DB) + \sim\!\sim\!\sim M_2\\
&\sim\!\sim\!\sim M_2^* + \sim\!\sim\!\sim M_2^* \rightarrow \sim\!\sim\!\sim M_2 + \sim\!\sim\!\sim M_2 (DB)
\end{aligned} \tag{4}$$

where (DB) indicates a terminal double bond has been formed. Likewise the reactions for transfer are

$$\begin{aligned}
&\sim\!\sim\!\sim M_1^* + M_1 \rightarrow \sim\!\sim\!\sim M_1 + M_1^*\\
&\sim\!\sim\!\sim M_1^* + M_2 \rightarrow \sim\!\sim\!\sim M_1 + M_2^*\\
&\sim\!\sim\!\sim M_2^* + M_1 \rightarrow \sim\!\sim\!\sim M_2 + M_1^*\\
&\sim\!\sim\!\sim M_2^* + M_2 \rightarrow \sim\!\sim\!\sim M_2 + M_2^*
\end{aligned} \tag{5}$$

The large number of termination and transfer constants precludes an exact analysis of the average molecular weights, especially because of the difficulty in determining the cross-termination and cross-transfer constants. Suffice it to say that a copolymer is produced with a number-average degree of polymerization of $\bar{r}_n$.

Stockmayer defines the average composition of M_1 in the copolymer chains as p_o. Then

$$p_o = \frac{M_1(r_1M_1 + M_2)}{r_1M_1^2 + 2M_1M_2 + r_2M_2^2} = 1 - q_o \tag{6}$$

where q_o is the average composition of M_2 in the copolymer. We now define a deviation function y:

$$y = p - p_o = q_o - q \tag{7}$$

where $\underline{p}$ is the composition of M_1 in an individual chain, independent of molecular weight. The weight fraction of copolymer molecules with lengths between $\underline{r}$ and $\underline{r} + \underline{dr}$ with a deviation in composition between $\underline{y}$ and $\underline{dy}$ is given by

$$W(r, y)\, dr\, dy = (1/\pi\phi)^{\frac{1}{2}} (r^{3/2}/\bar{r}_n^{\,2})$$

$$\times \exp [(-r/\bar{r}_n)(1 + \bar{r}_n y^2/\phi)]\, dr\, dy \tag{8}$$

(Stockmayer, 19) where

$$\phi = 2p_oq_o\kappa \tag{9}$$

$$\varkappa = [1 - 4p_o q_o (1 - r_1 r_2)]^{\frac{1}{2}} \tag{10}$$

$$= \frac{r_1 M_1^2 + 2r_1 r_2 M_1 M_2 + r_2 M_2^2}{r_1 M_1^2 + 2M_1 M_2 + r_2 M_2^2}$$

The weight fraction molecular weight distribution is given by

$$W(r)\ dr = \int_{-\infty}^{+\infty} [W(r, y)\ dr]\ dy$$

$$= (r/\bar{r}_n^{\,2}) \exp(-r/\bar{r}_n)\ dr \tag{11}$$

which is shown in Figure 1.2.2.

The composition distribution, irrespective of molecular weight, is

$$W(y)\ dy = \int_0^{\infty} [W(r, y)\ dy]\ dr$$

$$= \frac{3(\bar{r}_n/\phi)^{\frac{1}{2}}\ dy}{r[1 + (\bar{r}_n/\phi)y^2]^{5/2}} \tag{12}$$

(Stockmayer, 19).

Figure 2.11.1 shows W(y) against $p = y + p_o$ for two values of $\bar{r}_n/\phi$, that is, for the conditions of $M_1 = M_2$, $r_1 = 0.5$, $r_2 = 2.00$, for which $p_o = 0.333$ and with $\bar{r}_n = 100$ and 400. As seen in the figure, larger values of the parameter $\bar{r}_n/\phi$ cause the composition distribution

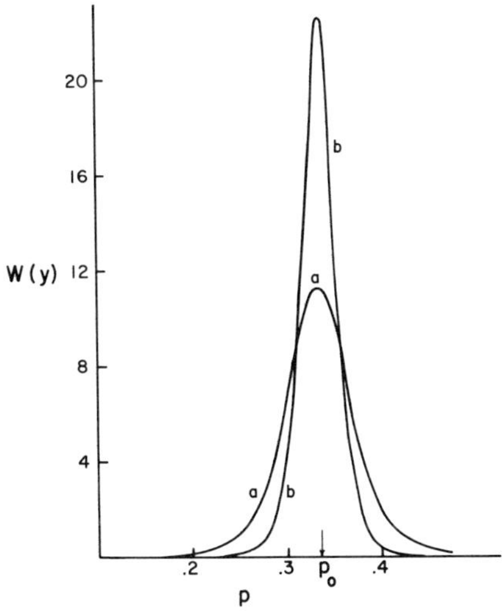

Figure 2.11.1 - Composition distribution of a copolymer as a function of the composition of an individual chain, independent of molecular weight, for $\bar{r}_n/\phi = 225$(a) and 900(b), and $p_o = 0.333$. (Courtesy *J. Chem. Phys.*) (Stockmayer, 19).

to become narrower. For the two cases selected, rather low values of $\bar{r}_n$ were used for free-radical type copolymers, hence for commercial solid copolymers, quite narrow distributions will be encountered. The ordinate scale in Figure 2.11.1 is considerably larger than those encountered in previous figures; the weight fraction of material between y and $y + dy$ is given by W(y) dy so for values of dy of 0.001, the weight fraction at the maximum is 22.5 (0.001) = 0.225.

12. Diffusion-Controlled Termination

A. Constant rate of initiation, monomer concentration varies, no transfer, termination by disproportionation and influenced by molecular diffusion.

In the later stages of some polymerizations, there is a rapid increase in the rate of polymerization, known as the Trommsdorff effect or gel effect. The rapid increase is attributed to a decrease in the rate of termination caused by a decrease in the diffusivity of the macroradicals in the increasingly viscous medium. If a specific rate constant, k_D, is dependent on its diffusivity, then we write

$$k_D = BD \tag{1}$$

where D is the diffusion constant and B is a proportionality factor. Ito assumes that the termination constant between radicals of size r and size s, $k_{t,rs}$ is

$$k_{t,rs} = BD_1(r^{-\frac{1}{2}} + s^{-\frac{1}{2}}) \tag{2}$$

where D_1 is the diffusion constant for monomer. He then uses the Kolomogorov "forward differential equation with termination" and numerical solution of an integral equation to find

$$F(r) = 10.8h^2 \exp\{-5.4h^2r - 2hr^{\frac{1}{2}} + 0.18\} \tag{3}$$

(Ito, 20) where

$$h = BD_1 \Sigma P_r / k_p M$$

and ΣP_r is the concentration of polymer = $P_r/F(r)$. Ito states that F(r) is normalized so that

$$\int_1^\infty F(r)\ dr = 1 \qquad (4)$$

but computer calculations for $\bar{r}_n$ = 100 indicate that the preexponential factor should be 8.93 instead of 10.8. Ito notes that the distribution probably does not apply to a real system because (a) if the termination rate becomes sufficiently slow, the steady-state assumption is no longer valid, (b) the equation for $k_{t,rs}$ assumes that all radicals are diffusion-controlled, and (c) the entire radical is required to diffuse, whereas only the active end needs to move. Comparison of the weight distribution function with the Schulz-Flory function shows that the former distribution is biased toward the high molecular weight polymers, Figure 2.12.1.

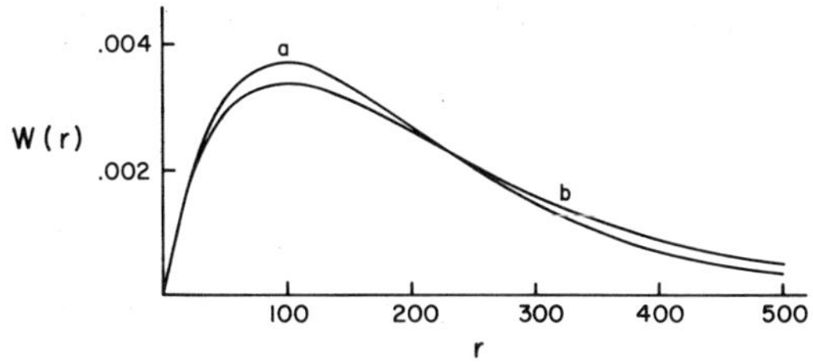

Figure 2.12.1 - Diffusion Controlled termination by disproportionation where the termination constant depends

on the size of the interacting species, (a). Compared with the Schulz-Flory distribution for $\bar{r}_n = 100$, (b) (after Ito, 20).

13. Primary Radical Termination

A. Constant rate of initiation, monomer concentration varies, transfer to monomer and to solvent, termination either by second-order disproportionation or by second-order combination and by interaction with primary radicals

When high rates of initiation are used, the primary radicals so generated not only initiate chains but also can terminate growing chains. This process is called "primary radical termination." The kinetic features of polymerization which incorporates primary radical termination are discussed by Bamford, Jenkins, and Johnson (21) and by Bevington (22). The average degrees of polymerization can be obtained by use of the method described in section 5.3 below. Specifically, if we assume a constant rate of initiation, invariant monomer concentration, transfer to monomer, transfer to solvent, and second-order termination by primary radicals and by either disproportionation or combination, then we can use the steady-state assumption. In practical cases, the steady-state assumption may not be valid; but at least we can study the effect of primary radical termination under steady state conditions. This assumption requires

$$dY_n/dt = d\Sigma r^n R_r^*/dt = 0 \qquad (1)$$

hence

$$Y_o = \left\{ \frac{-k_{prt}R_I}{k_iM} + \left[\left(\frac{k_{prt}R_I}{k_iM} \right)^2 + 2k_{t,d}R_I \right]^{\frac{1}{2}} \right\} / 2k_{t,d} \tag{2}$$

$$Y_1 = \frac{R_I + (k_pM + k_{tr,m}M + k_{tr,s}S)Y_o}{[2k_{t,d}Y_o + (2k_{prt}R_I/k_iM) + k_{tr,m}M + k_{tr,s}S]} \tag{3}$$

$$Y_2 = \frac{R_I + 2k_pMY_1 + (k_pM + k_{tr,m}M + k_{tr,s}S)Y_o}{[2k_{t,d}Y_o + (2k_{prt}R_I/k_iM) + k_{tr,m}M + k_{tr,s}S]} \tag{4}$$

for termination by disproportionation, and similar expressions with $k_{t,d}$ replaced by $k_{t,c}$ for termination by combination. The specific rate constant for primary radical termination is k_{prt}

$$R_r^* + I^* \rightarrow P_r \qquad k_{prt} \tag{5}$$

and we use the convention

$$dP_r/dt = 2k_{prt}R_r^*I^* + \ldots \tag{6}$$

The rate of initiation by I^* radicals is

$$R_I = k_iI^*M \tag{7}$$

For termination by disproportionation

$$\bar{r}_n = 1 + k_pM/[(R_I/Y_o) + k_{tr,s}S + k_{tr,m}M] \tag{8}$$

$$\bar{r}_w = 1 + 2k_pM/[(R_I/Y_o) + k_{tr,m}M + k_{tr,s}S] \tag{9}$$

hence, there is essentially no change in the distribution because, for long chains, $\bar{r}_w/\bar{r}_n = 2$ with or without primary radical termination. The situation is more complex for termination by combination

$$\bar{r}_n = \frac{R_I + (k_pM + k_{tr,m}M + k_{tr,s}S)Y_o}{R_I/2 + [(k_{prt}R_I/k_iM) + k_{tr,m}M + k_{tr,s}S]Y_o} \tag{10}$$

$$\bar{r}_w = 1 + \frac{2k_{t,c}Y_1^2 + 2k_pMY_1}{R_I + (k_pM + k_{tr,m}M + k_{tr,s}S)Y_o} \tag{11}$$

To find $\bar{r}_n$ and $\bar{r}_w$ as a function of conversion when termination is by combination, use the procedure given in section 2.4 and numerical integration.

14. References

1. W. E. Davis as quoted in J. C. W. Chien, J. Am. Chem. Soc., 81, 86 (1959).
2. C. H. Bamford, "On Free Radical Polymerizations with Rapidly Decaying Initiators," Polymer, 6, 63 (1965).
3. J. C. W. Chien, "Kinetics of Ethylene Polymerization Catalyzed by Bis-(cyclopentadienyl)-titanium Dichloride-Dimethylaluminum Chloride," J. Am. Chem. Soc., 81, 86 (1959).
4. L. Küchler, Polymerisationskinetik, Springer-Verlag, Berlin, 1951.

5. C. H. Bamford, W. G. Barb, A. D. Jenkins, and P. F. Onyon, The Kinetics of Vinyl Polymerization by Radical Means, Academic Press, London, 1958.
6. P. J. Flory, Principles of Polymer Chemistry, Cornell University Press, Ithaca, N. Y., 1953.
7. G. V. Schulz, "Uber die Kinetik der Kettenpolymerisationen. V. Der Einfluss verschiedener Reaktionsarten auf die Polymolekularität," Z. Physik. Chem., B43, 25 (1939).
8. See, for example, J. B. Dale, Five-Figure Tables of Mathematical Functions, Arnold, London, 1949, p. 112; E. Jahnke and F. Emde, Tables of Functions with Formulae and Curves, Dover Publications, 1945, New York; C. R. C. Standard Mathematical Tables, Chemical Rubber Publishing Co., Cleveland, 1960.
9. A. V. Tobolsky, R. H. Hobran, R. Böhme, and R. Schaffhauser, "Heterogeneity Index During Dead-End Polymerization," J. Phys. Chem., 67, 2336 (1963).
10. D. O. Jordan and A. R. Mathieson, "The Kinetics of Catalytic Polymerizations. IV. Molecular-Weight Distribution in Polar Polymerizations," J. Chem. Soc., 2358 (1952). (Note: equation 13, et seq. contain errors owing to an integration error.)
11. M. Litt, "Molecular Weights in Cationic Polymerization," J. Polymer Sci., 43, 567 (1960).
12. G. V. Schulz, "Uber die Beziehung zwischen Reaktionsgeschwindigkeit und Zusammensetzung des Reaktionsprodukte bei Makropolymerisationsvorgängen," Z. Physik. Chem., B30, 379 (1935).

13. C. H. Bamford and H. Tompa, "Calculation of Molecular Weight Distributions from Kinetic Schemes," Trans. Faraday Soc., 50, 1097 (1954).

14. J. C. W. Chien, "Olefin Polymerizations and Polyolefin Molecular Weight Distribution," J. Polymer Sci., A1, 1839 (1963).

15. A. V. Tobolsky, "Dead-End Radical Polymerization," J. Am. Chem. Soc., 80, 5927 (1958); A. V. Tobolsky, C. E. Rogers, and R. D. Brinkman, "Dead-End Radical Polymerization. II," J. Am. Chem. Soc., 82, 1277 (1960).

16. G. E. Ham, Editor, Copolymerization, Interscience, New York, 1964.

17. J. Brandrup and E. H. Immergut, Polymer Handbook, Interscience, New York, 1966.

18. T. Alfrey, Jr., J. J. Bohrer, and H. Mark, Copolymerization, Interscience, New York, 1952.

19. W. H. Stockmayer, "Distribution of Chain Lengths and Compositions in Copolymers," J. Chem. Phys., 13, 199 (1945).

20. K. Ito, "Derivation of Distributions of the Degree of Polymerization by Probability Theory," J. Polymer Sci., A2, 7, 241 (1969).

21. C. H. Bamford, A. D. Jenkins, and R. Johnson, "Termination by Primary Radicals in Vinyl Polymerization," Trans. Faraday Soc., 55, 1451 (1959).

22. J. C. Bevington, Radical Polymerization, Academic Press, New York, 1961.

Chapter 3

Addition Polymerization--"Living Polymers with Partial Deactivation"

Contents

1. Introduction and the Poisson Distribution

Rate of initiation equals $k_p MI$, monomer concentration varies, no transfer, no termination.

Flory (1) described the distribution produced when an initiator such as alcohol is added to a cyclic monomer such as ethylene oxide. Because there are no termination reactions, the number-average degree of polymerization in this system is the number of initial monomer molecules divided by the number of initiator molecules,

$$\bar{r}_n = M_o/I_o \tag{1}$$

provided that the process of mixing monomer and initiator is faster than the process of polymerization. Now, if we count the initiator as a unit in the chain, then

$$\bar{r}_n = 1 + M_o/I_o = 1 + \nu \tag{2}$$

For this condition, the frequency distribution is the Poisson distribution

$$F(r) = \exp(-\nu)\nu^{(r-1)}/(r - 1)! \tag{3}$$

The weight distribution is given by

$$W(r) = r \exp(-\nu)\nu^{(r-1)}/(r - 1)!(\nu + 1) \tag{4}$$

(Flory, 1). For all practical purposes, one cannot distinguish between F(r) and W(r), even though W(r) is slightly skewed toward larger values of r. The variation of $\bar{r}_n$ with time is

$$\bar{r}_n = 1 + \frac{(M_o - M)}{(I_o - I)} = 1 + \nu \qquad (5)$$

$$\bar{r}_w = 1 + \nu + \nu/(1 + \nu) \approx 1 + \bar{r}_n \qquad (6)$$

Curves for F(r) and W(r) versus r are given in Figures 1.7.1, 1.7.2, and 3.1.1.

If each initiator molecule contains two active centers, then ν is defined by

$$\nu = 2(M_o - M)/(I_o - I) \qquad (7)$$

The distribution functions and the average degrees of polymerization are given by equations (3) to (6).

The distributions which are produced by polymerization without termination are considerably narrower than can be achieved by fractionation of a broadly distributed polymer, such as that discussed in Chapter 2. With the advent of anionic initiators under extremely rigorous conditions of purity, vinyl polymers can be prepared with Poisson distributions. Thus it is possible, in principle, to manufacture large amounts of "nearly monodisperse" polymers and to investigate the influence of molecular weight distribution on a polymer while holding the number-average molecular weight constant. In practice, high molecular weight polymers which have the Poisson distribution are

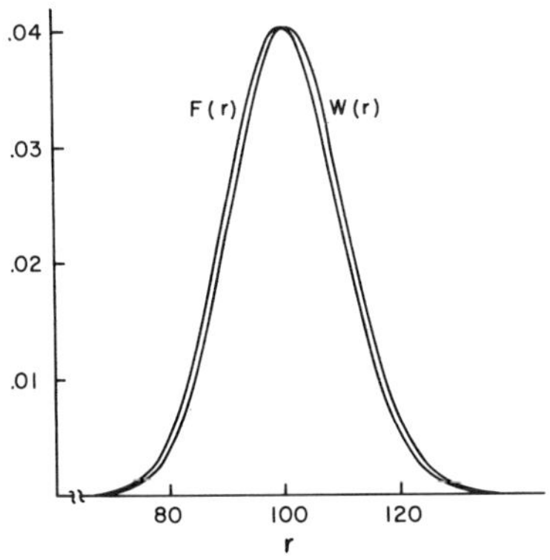

Figure 3.1.1 - Comparison of frequency and weight distribution as a function of $\underline{r}$ for a "Poisson" type polymer when $\bar{r}_n = 100$.

difficult to achieve, primarily because the active polymers become deactivated by transfer, termination, or reaction with an impurity. If all the chains become deactivated, the distributions of Chapter 2 may result; hence we concern ourselves in this chapter primarily with partial deactivation.

2. The Gold Distribution

Rate of initiation equals k_iMI, $k_i \neq k_p$, monomer concentration varies, no transfer, no termination.

Section 1 gives the results for this type of polymerization when $k_i = k_p$. We consider here the general case when $k_i \neq k_p$, known as the Gold distribution (2). High molecular weight polymers with a Poisson distribution are difficult to achieve in practice; one reason may be that initiation is rather slow, leading to more broadly distributed polymers. The variation of unreacted initiator concentration with time is

$$I = -\int_0^t k_i IM \, dt \tag{1}$$

From dM/dt and dI/dt we find

$$M = M_o + \gamma I_o \ln (I/I_o) + (1 - \gamma)(I - I_o) \tag{2}$$

where

$$\gamma = k_p/k_i \tag{3}$$

In general the fraction of unreacted initiator, I/I_o, is very small, 10^{-3} or less, for high molecular weight polymer at conversions above 10% unless γ is very large, 10^3 or greater. The weight distribution function is

$$W(r) = \frac{r}{\gamma \bar{r}_n (1 - I/I_o)} \left(\frac{I}{I_o}\right) \left(\frac{\gamma}{\gamma - 1}\right)^r$$

$$\times \left[1 - \left(\frac{I}{I_o}\right)^{\gamma-1} \sum_{s=0}^{r-1} [(1 - \gamma) \ln (I/I_o)]^s/s!\right] \tag{4}$$

(Gold, 2).

If $\gamma > 1$, we can use the identity

$$Q(\chi^2|\nu) = \sum_{j=0}^{r-1} e^{-m} m^j/j! \tag{5}$$

where $r = \nu/2$, $m = \chi^2/2$, ν is even, and $1 - Q(\chi^2 \mid \nu)$ is the chi-square probability function. Use can also be made of the approximation

$$Q(\chi^2 \mid \nu) \approx Q(x) \tag{6}$$

$$x = (m - r)/r^{\frac{1}{2}}, \qquad r > 15$$

where

$$Q(x) = (2\pi)^{-\frac{1}{2}} \int_x^{\infty} \exp(-t^2/2)\, dt \tag{7}$$

The latter is found in many statistical handbooks. If $\gamma < 1$, the sum term cannot be easily approximated, and the term $u = (1 - \gamma) \ln I/I_o$ is a large negative number, 10^2, 10^3, or so. In general as γ becomes small, less than 1, the distribution does not differ significantly from the Poisson distribution. This is easily understood by considering the reactions

$$I^* + M \xrightarrow{k_i} R_2^*$$

$$R_n^* + M \xrightarrow{k_p} R_{n+1}^*$$

As k_i becomes equal to or greater than k_p, the rate-controlling step is propagation. Thus instantaneous initiation will produce a Poisson distribution except for very low molecular weights, $\bar{r}_n$ in the range of 2 to 10. Instantaneous initiation is used in subsequent cases to simplify the mathematical treatment. The number and weight distribution functions are shown in Figures 3.2.1 through 3.2.3 for $\bar{r}_n = 100$, as the parameter γ varies from 10^0 to 10^4.

Curve	b	c	d	e
γ	10^0	10^2	10^3	10^4

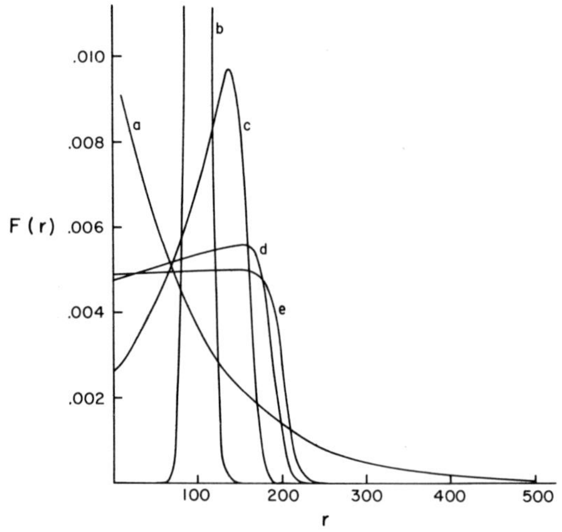

Figure 3.2.1 - Frequency distribution as a function of r for "living polymers" without termination, $\gamma = k_p/k_i$ as indicated, $\bar{r}_n = 100$. The Schulz-Flory distribution is also shown for $\bar{r}_n = 100$. (a) (after Gold, 2).

Note that for very large γ, the frequency distribution approximates a step function, that is,

$$F(r) = \text{constant}, \quad r \leq r_{max}$$

Curve	b	c	d	e	f
γ	10^0	10^1	10^2	10^3	10^4

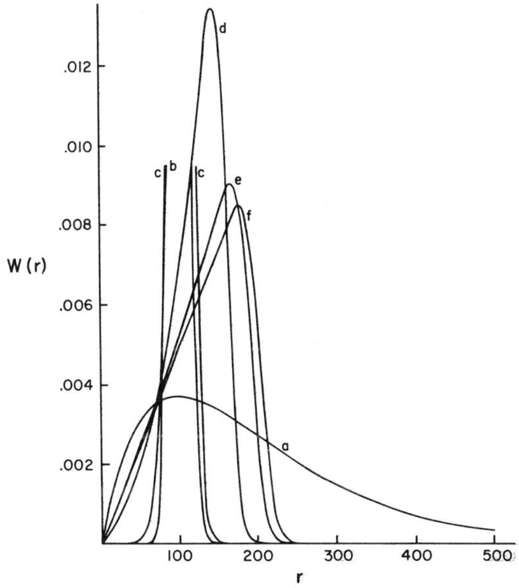

Figure 3.2.2 - Weight distribution as a function of r. Same conditions as in Figure 3.2.1. Schulz-Flory distribution (a).

$$F(r) = 0 \qquad r \geqslant r_{max}$$

where "constant" = $1/r_{max}$

$$\bar{r}_n = \frac{M_o - M}{I_o - I} \qquad (8)$$

$$= [(1 - \gamma)(1 - e^{-u/R}) + \gamma u/R]/(1 - e^{-u/R}) \qquad (9)$$

where $R = |\gamma - 1|$, $\gamma \neq 1$, and $u = R \ln(I/I_o)$

Curve	a	b	c	d	e
γ	10^0	10^1	10^2	10^3	10^4

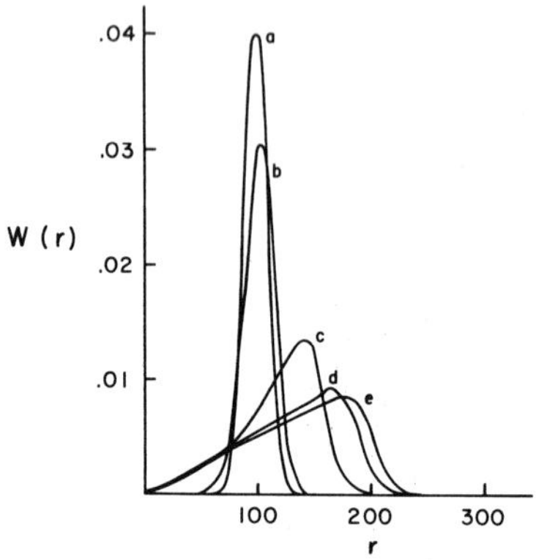

Figure 3.2.3 - Figure 3.2.2 redrawn to show the entire curves.

$$\bar{r}_w = [(\gamma u/R)(1 + \gamma u/R) - 2\gamma u$$

$$+ (2\gamma - 1)(\gamma - 1)(1 - e^{-u/R})]/[(1 - \gamma)(1 - e^{-u/R})$$

$$+ (\gamma u/R)] \tag{10}$$

$$= [(2\gamma - 1) - (2\gamma - 3)(\bar{r}_n + \gamma - 1)$$

$$+ (\bar{r}_n + \gamma - 1)^2 (1 - I/I_o)]/(\bar{r}_n) \qquad \gamma > 1 \tag{11}$$

The $\bar{r}_n$, $\bar{r}_w$, and $\bar{r}_w/\bar{r}_n$ are given in Figures 3.2.4 and 3.2.5 and as a function of _u_. The $\bar{r}_w/\bar{r}_n$ goes through a maximum as _u_ increases, then approaching 1.00 as _u_ becomes even larger (Figures 3.2.6 and 3.2.7). For moderate values of _u_, 10^2 or less, the maximum in $\bar{r}_w/\bar{r}_n$ occurs at $(M_o - M)/(I_o - I) \approx 1$ or at very low conversions.

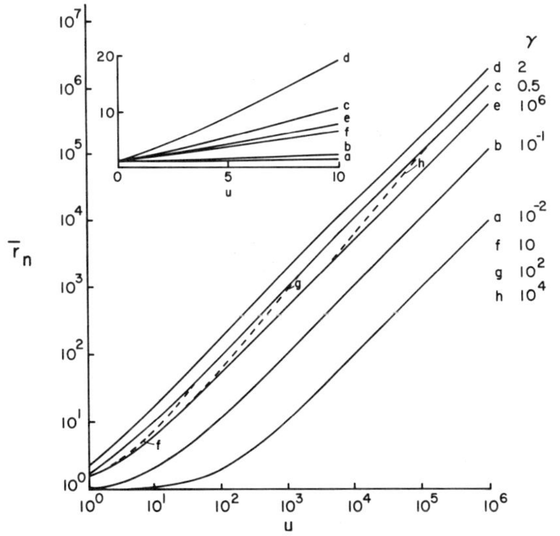

Figure 3.2.4 - Number average degree of polymerization as a function of the parameter $u = -|\gamma - 1| \ln(I/I_o)$ for the Gold distribution, no termination, and $\gamma = k_p/k_i$. The main figure is plotted versus log u while the insert gives $\bar{r}_n$ versus u. (Courtesy J. Chem. Phys.) (Gold, 2).

However, for very large values of γ, M is exhausted before much of the catalyst has reacted; hence broad distributions result.

3. Partial Deactivation by a First-Order Process or a Second-Order Process with an Impurity

Instantaneous initiation, monomer concentration may vary, no transfer, termination as specified.

A. The general case.

We want to consider the effect of a small amount of

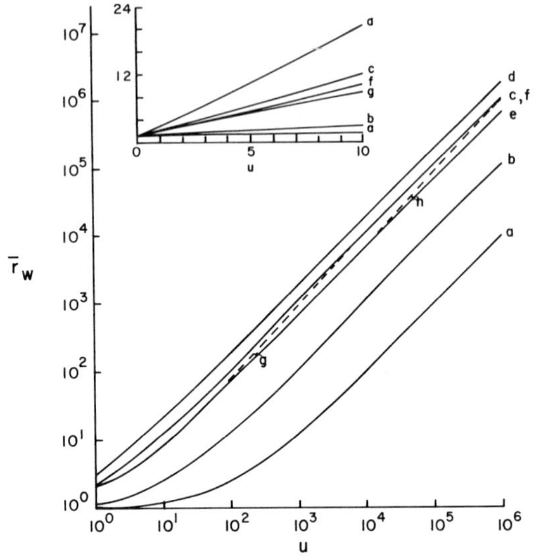

Figure 3.2.5 - Weight average degree of polymerization as a function of u. Same conditions as in Figure 3.2.4. (Courtesy J. Chem. Phys.) (Gold, 2).

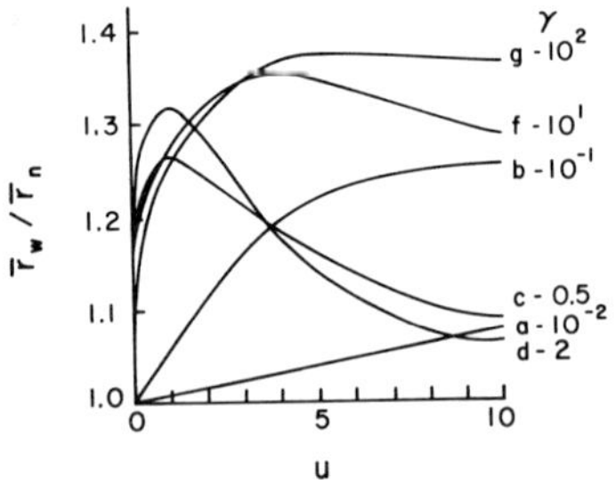

Figure 3.2.6 - Variation of $\bar{r}_w/\bar{r}_n$ as a function of u for the Gold distribution. (Courtesy J. Chem. Phys.) (Gold, 2).

termination upon the "living polymer" distribution; that is, living polymers with partial deactivation. Let us

Curve	a	b	c	d	e	f	g	h
γ	10^{-2}	10^{-1}	0.5	2	10^{6}	10^{1}	10^{2}	10^{4}

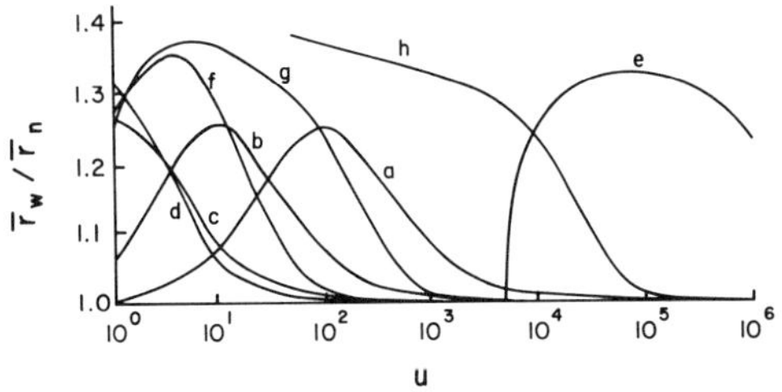

Figure 3.2.7 - Variation of $\bar{r}_w/\bar{r}_n$ as a function of log u. For values of $\gamma < 10^2$, the maximum occurs at $\bar{r}_n \approx 1$ or at very low conversions. (Courtesy J. Chem. Phys.) (Gold,2).

assume that deactivation is either a first-order process or a second-order process with an impurity:

$$R_n^* \xrightarrow{k_t} P_n$$

$$R_n^* + S \xrightarrow{k_{t,s}} P_n \qquad (1)$$

The constraint is made such that the total number of molecules present, not counting monomer, is equal to the initial number of initiator molecules:

$$\Sigma R_n^* + \Sigma P_n = R_o^* \qquad (2)$$

We define the parameter ψ such that

$$\psi = \int_0^t k_t \, d\tau \qquad (3)$$

if termination is by a first-order reaction.

$$\psi = \int_0^t k_{t,s} S \, d\tau \tag{4}$$

if termination is by a second-order reaction with solvent or impurity.

With this definition, ψ, k_t, $k_{t,s}$, and $\underline{S}$ can vary with time, but they are independent of the size of the molecule $\underline{r}$.

Thus

$$\Sigma R_n^* = R_o^* \exp(-\psi) \tag{5}$$

We also define the parameter θ

$$\theta = \int_0^t k_p M \, d\tau \tag{6}$$

With these definitions, the mole fraction of active radicals of size $\underline{r}$ is

$$F(r)(\text{active}) = \{\theta^{r-1} \exp[-(\psi + \theta)]\}/(r - 1)! \tag{7}$$

The mole fraction of dead polymer molecules is

$$F(r)(\text{dead}) = \int_0^t \frac{\theta^{r-1} \exp[-(\psi + \theta)]}{(r-1)!} k_t \, d\tau \tag{8}$$

$$= \int_0^t \frac{\theta^{r-1} \exp[-(\psi + \theta)]}{(r - 1)!} k_{t,s} S \, d\tau \tag{9}$$

(Coleman, Gornick, and Weiss, 3) depending on the mode of termination.

$$\bar{r}_n = 1 + \int_0^t k_p M \exp(-\psi)\, d\tau \tag{10}$$

$$\bar{r}_w = [1 + 3 \int_0^t k_p M \exp(-\psi)\, d\tau + 2 \int_0^t \theta k_p M \exp(-\psi)\, d\tau] \div (\bar{r}_n) \tag{11}$$

B. Rate of termination/rate of propagation is a constant.

Let $d\psi/d\theta = \delta$ = a constant. Thus $\psi/\theta = \delta$, or $S/(S + M)$ = a constant. This is a restrictive situation as the solvent or impurity must be used up at the same rate as the monomer. For this case

$$F(r)(\text{active}) = \theta^{r-1} \{\exp[-\theta(1 + \delta)]\}/(r - 1)!$$

$$F(r)(\text{dead}) = \delta \int_0^\theta \frac{\theta^{r-1} \exp[-\theta(1 + \delta)]}{(r - 1)!}\, d\theta \tag{12}$$

(Coleman, Gornick, and Weiss, 3). Subject to the condition

$$\sum_{r=1}^{\infty} F(r)(\text{active}) + \sum_{r=1}^{\infty} F(r)(\text{dead}) = 1 \tag{13}$$

$$\bar{r}_n = 1 + [1 - \exp(-\delta\theta)]/\delta \tag{14}$$

$$\bar{r}_w = \{1 + 3[1 - \exp(-\delta\theta)]/\delta + 2[1 - \exp(-\delta\theta)]/\delta^2$$

$$- 2\theta[\exp(-\delta\theta)]/\delta\}/\bar{r}_n \quad (15)$$

Orofino and Wenger (4) have considered this particular case in more detail. For a monofunctional initiator, the weight distribution is given by

$$W(r) = [r(1 - \delta)^{r-2}/\bar{r}_n][\delta(1 - \Phi)$$

$$+ (1 - \delta)(\theta - 1)^{r-1} \exp(1 - \theta)/(r - 1)!] \quad (16)$$

(Orofino and Wenger, 4) where

$$\Phi(t) = (1/2\pi)^{\frac{1}{2}} \int_{-\infty}^{t} \exp(-y^2/2)\, dy \quad (17)$$

$$t = (r - \theta)/(\theta - 1)^{\frac{1}{2}} \quad (18)$$

$$\theta = 1 - (1/\delta) \ln (1 + \delta - \delta\bar{r}_n) \quad (19)$$

The form of W(r) appears different from equations (12) and (13) because of the method of counting and because a summation is replaced by the normal approximation to the Poisson distribution.

This function is plotted for $\bar{r}_n = 101$ for various δ's in Figure 3.3.1. Note that under the specified conditions of $\bar{r}_n = 101$, a 1% impurity leads to a Schulz-Flory type distribution, which results from bimolecular termination either by disproportionation or by transfer. They also show that the maximum possible molecular weight is

$$\bar{r}_n \max = (1 + \delta)/\delta \tag{20}$$

which is 101 for 1% impurity. Letting $p = 1/(1 + \delta)$, we find

$$\bar{r}_n \max = 1/(1 - p) \tag{21}$$

$$\bar{r}_w \max = (1 + p)/(1 - p) \tag{22}$$

which are the Schulz-Flory equations.

The fraction of chains which are deactivated is

$$\beta_I = \delta(\bar{r}_n - 1) \tag{23}$$

for monoactive polymerization

$$\bar{r}_w = 1 + 2/\delta - 2/\delta\bar{r}_n$$

$$+ (2/\bar{r}_n)(1 - 1/\delta)(\bar{r}_n - 1 - 1/\delta)\ \ln(1 + \delta - \delta\bar{r}_n) \tag{24}$$

If a diactive initiator is used, then

$$W(r) = [r(1 - \delta)^{r-4}/\bar{r}_n]\{(\theta - 2)^{r-2}(1 - \delta)^2$$

$$\times \exp(2 - \theta)/(r - 2)! + \delta^2[1 - \Phi[(r - \theta)/(\theta - 2)^{\frac{1}{2}}]]$$

$$+ 2\delta(1 - \delta) \sum_{j=0}^{r-3} (\theta/2 - 1)^j \exp(1 - \theta/2)/j!]$$

$$\times [1 - \Phi[(r - j - \theta/2 - 1)/(\theta/2 - 1)^{\frac{1}{2}}]]\} \qquad (25)$$

(Orofino and Wenger, 4) where

$$\theta = 2[1 - (1/\delta)\ \ln\ (1 + \delta - \delta\,\bar{r}_n/2)] \qquad (26)$$

Comparison of the W(r) for mono- and diactive molecules for $\delta = 0$ and 0.5% is given in Figure 3.3.2. Because each molecule of the dianion has two active ends, the distribution does not depart as much from the original. At high values of δ, the distribution approaches a Schulz distribution with k = 2; that is, bimolecular termination by combination.

The maximum molecular weight for diactive polymerization is

$$\bar{r}_n\ \text{max} = 2(1 + \delta)/\delta \qquad (27)$$

while the fraction of chains deactivated at one or both ends is

$$\beta_{II} = 1 - (1 + \delta - \delta\bar{r}_n/2)^2 \qquad (28)$$

In general, $\bar{r}_w$ for diactive polymerization is given by

$$\bar{r}_w = (1/\delta^2\bar{r}_n)[\delta^2\bar{r}_n(1 + \bar{r}_n/2) + 2\delta(\bar{r}_n - 2)$$

$$+ 4\delta(1/\delta - 1)(1 + \delta - \delta\bar{r}_n/2)\ \ln(1 + \delta - \delta\bar{r}_n/2)] \qquad (29)$$

Curve	a	b	c	d	e
δ	0	0.003	0.005	0.008	0.010
$\bar{r}_w/\bar{r}_n$	1.01	1.12	1.23	1.48	1.97

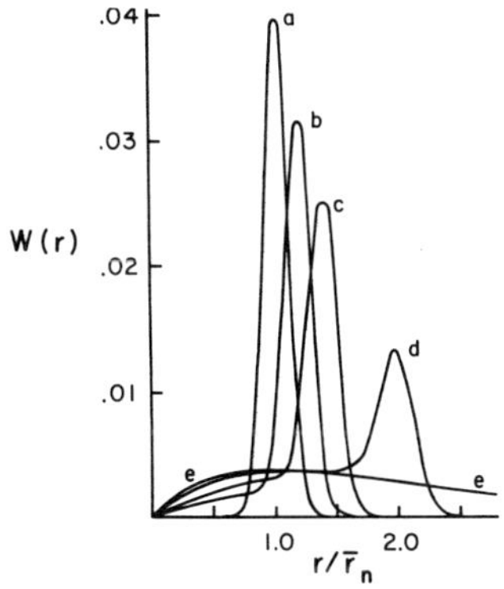

Figure 3.3.1 - Weight distribution as a function of $r/\bar{r}_n$ for "living polymers" in which the ratio of the rate of deactivation to the rate of propagation always remains equal to a constant, δ, and $\bar{r}_n = 101$. (Courtesy J. Chem. Phys.) (Orofino and Wenger, 4).

Curve	a	b	c	d
δ	0	0.005	0.005	0.02
$\bar{r}_w/\bar{r}_n$	1.01	1.05	1.23	1.44

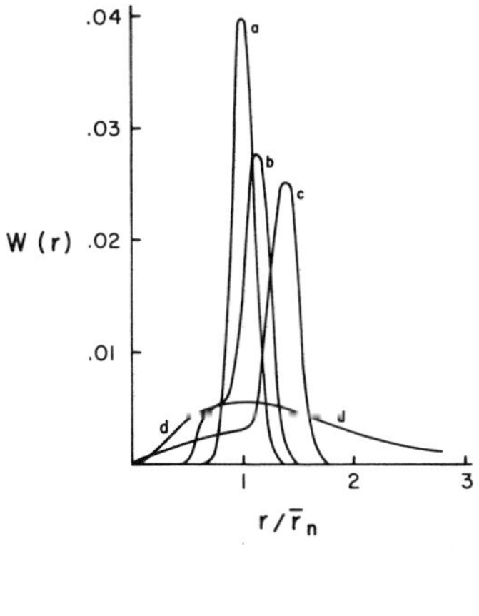

Figure 3.3.2 - Weight distribution as a function of $r/\bar{r}_n$ for monoactive (a, c) and diactive (a, b, d) initiators in a "living system" described in Figure 3.3.1. (Courtesy J. Chem. Phys.) (Orofino and Wenger, 4).

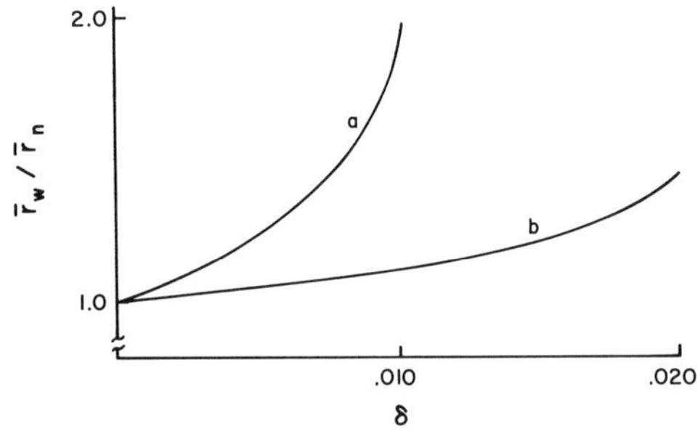

Figure 3.3.3 - Variation of the dispersion ratio $\bar{r}_w/\bar{r}_n$ as a function of the parameter δ for monoactive (a), (equation 24) and diactive (b), (equation 29) initiators. (after Orofino and Wenger, 4).

The dispersion ratio, $\bar{r}_w/\bar{r}_n$, is shown in Figure 3.3.3 as a function of δ for monoactive (Equation 23) and diactive (Equation 29) initiators with $\bar{r}_n = 100$.

C. Termination by an impurity.

The more difficult problem of polymerization with a termination rate independent of the rate of polymerization is treated by Coleman, Gornick, and Weiss. Define the following dimensionless parameters

$$\alpha = M/M_o, \qquad \beta = S/S_o, \qquad \sigma = S_o/I_o, \qquad \delta = k_{t,s}/k_p$$

They show that

$$\beta = (1 - \sigma)/\{\exp[(1 - \sigma)k_{t,s}I_o t] - \sigma\} \tag{30}$$

$$\alpha = \beta^{(1/\delta)} \tag{31}$$

$$\psi = -\ln[1 - \sigma(1 - \alpha^{\delta})] \tag{32}$$

$$\theta = \frac{M_o}{I_o} \int_{\alpha}^{1} \frac{d\alpha}{[1 - \sigma(1 - \alpha^{\delta})]} \tag{33}$$

$$\bar{r}_n = 1 + M_o (1 - \alpha)/I_o \tag{34}$$

The equations for the weight-average degree of polymerization are quite complex. For the general case

$$\bar{r}_w/\bar{r}_n = 1 + 1/\bar{r}_n - 1/\bar{r}_n^{\,2}$$

$$+ (1 - 1/\bar{r}_n)^2 \sum_{j=1}^{\infty} \sigma^j C_j(\alpha, \delta) \tag{35}$$

where

$$C_j(\alpha, \delta) = \frac{2}{(1 - \alpha)^2} \sum_{k=1}^{j} (-1)^k \binom{j}{k} \left[\frac{1 - \alpha^{k\delta+2}}{k\delta + 2}\right.$$

$$\left. - \frac{\alpha(1 - \alpha^{k\delta + 1})}{k\delta + 1}\right] \tag{36}$$

Figure 3.3.4 shows how $\sum_{j=1}^{\infty} \sigma^j C_j(\alpha, \delta)$ varies with conversion when $\sigma = S_o/I_o = 1/5$ and $\delta = 1/2$ or 2. Thus, for

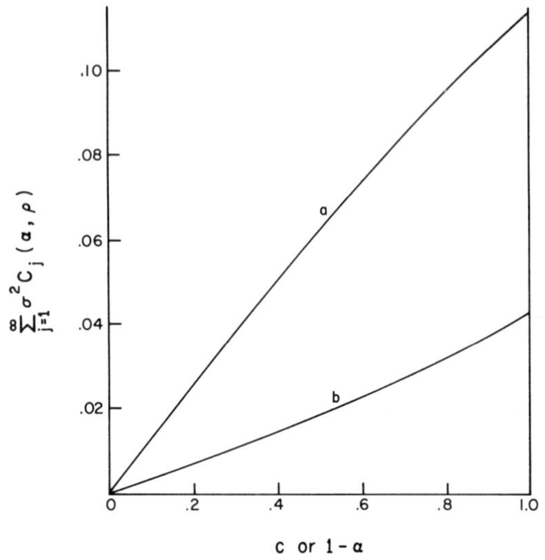

Figure 3.3.4 - Variation in the sum term in equation (35) as a function of conversion when $S_o/I_o = 1/5$ and $\delta = 2$(a) and 1/2(b). (Courtesy *J. Chem. Phys.*)(Coleman, Gornick, and Weiss, 3).

$\bar{r}_n = 100$, $\bar{r}_w/\bar{r}_n$ varies from 1.01 to 1.12 under these conditions as the conversion varies from 0 to unity. If σ is very small, so that most of the chains remain unterminated, only the first item in equation (36) is required; that is, $j = 1$ only. In this case

$$C_1(\alpha,\ \delta) = 1 - \frac{2}{(1 - \alpha)^2}\left[\frac{1 - \alpha^{\delta+2}}{\delta + 2} - \frac{\alpha(1 - \alpha^{\delta+1})}{\delta + 1}\right] \qquad (37)$$

The fraction of active chains is given by

$$I/I_o = 1 - \sigma(1 - \alpha^{\delta}) \qquad (38)$$

The fraction of active chains varies from unity at $\alpha = 1$ to $1 - \sigma$ at $\alpha = 0$. The variation of active chains with conversion is given in Figure 3.3.5 for the same values of σ and δ given in Figure 3.3.4.

D. First-order deactivation.

Chiang and Hermans (5) and also Guyot (6) have considered the situation with instantaneous initiation and when $t = \infty$, that is, when no active chains remain; compare equation (5). The final monomer concentration is

$$M_f = M_o \exp(-\eta I_o) \tag{39}$$

where

$$\eta = k_p/k_t$$

$$\bar{r}_n(t = \infty) = 1 + (M_o/I_o)\{1 - \exp(-\eta I_o)\} \tag{40}$$

$$\bar{r}_w(t = \infty) = \{1 + \frac{3M_o}{I_o}(1 - e^{-\eta I_o}) + \frac{2\eta M_o^2}{I_o} e^{-2\eta I_o}$$

$$\times [Ei(2\eta I_o) - Ei(\eta I_o) - \ln 2]\}/\{1 + \frac{M_o}{I_o}(1 - e^{-\eta I_o})\} \tag{41}$$

where $Ei(u) = \int_{-\infty}^{u} (e^y/y)\, dy$, the values of which are found in many mathematical handbooks (7).

Figure 3.3.6 shows $\bar{r}_w/\bar{r}_n$ as a function of η. At small η, k_t large relative to k_p, a most probable distribution is indicated, whereas as η increases, the distribution becomes sharper and becomes a Poisson distribution as

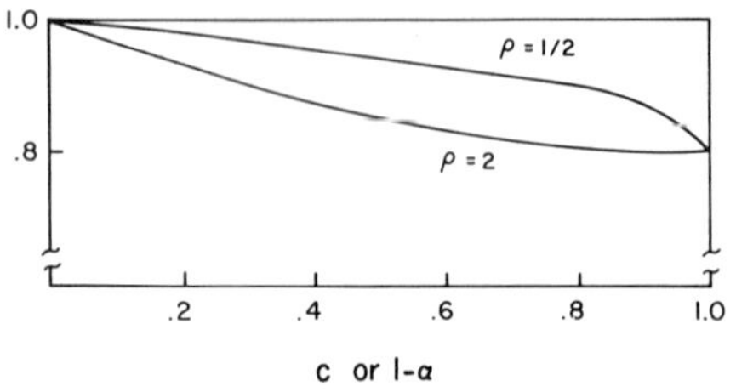

Figure 3.3.5 - Fraction of active chains remaining as a function of conversion. Same conditions as in Figure 3.3.4.

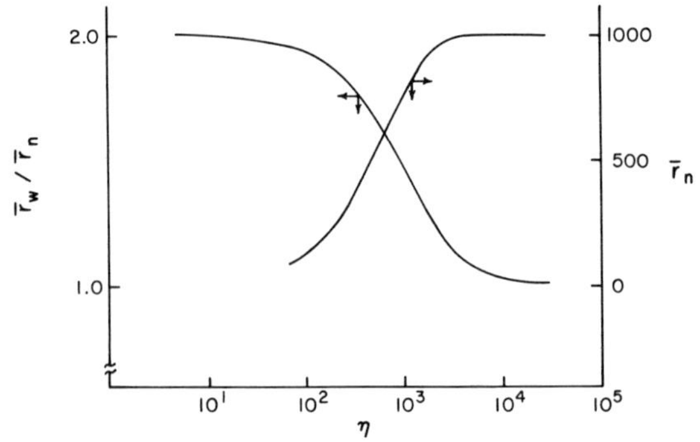

Figure 3.3.6 - Variation of the dispersion ratio $\bar{r}_w/\bar{r}_n$ and of $\bar{r}_n$ as a function of the ratio $\eta = k_p/k_t$ for $M_o/I_o = 1000$, $I_o = 1.504 \times 10^3$ moles/kg, instantaneous initiation, slow first-order deactivation. (after Chiang and Hermans, 5).

$\eta \to \infty$. In Chapter 2 we demonstrated that a "most probable" distribution resulted when we invoked first-order termination or transfer to monomer, regardless of the value of $\lambda = 1/\eta = k_t/k_p$; but in that derivation, the steady state assumption was made, that is, $dR^*/dt = 0$. This assumption is inconsistent with the reaction scheme used by Chiang and Hermans: here deactivation is a slow first-order reaction as specified.

4. Two Propagating Species

Rate of initiation equals $k_i MI^*$, monomer concentration varies, no transfer, no termination (8, 9).

To find the effect of two propagating species with different activities, let R_n^* be able to transform into a species Q_n^*

$$R_n^* \underset{k_{qr}}{\overset{k_{rq}}{\rightleftharpoons}} Q_n^* \tag{1}$$

This equation does not mean that R_n^* is in equilibrium with Q_n^*, rather that R_n^* can transform into Q_n^* and vice versa. There is an equilibrium condition that the total concentration of $R^* = \sum_{n=1}^{\infty} R_n^*$ is in equilibrium with the total concentration of

$$Q^* = \sum_{n=1}^{\infty} Q_n^*$$

$$Q^*/R^* = k_{rq}/k_{qr} = K \tag{2}$$

This is equivalent to stating that prior to addition of monomer, the active initiator is in equilibrium with inactive initiator; then it follows that the equilibrium condition remains after polymerization starts. Let both R_n^* and Q_n^* add monomer with rate constants k_r and k_q, respectively, then

$$M = M_o\{\exp -(k_r R^* + k_q Q^*)t\} \tag{3}$$

$$\bar{r}_n = 1 + (M_o - M)/I_o \tag{4}$$

the initiator $\underline{I}$ is counted in $\bar{r}_n$. The distribution function is difficult to evaluate, however for $M_o >> I_o$

$$\bar{r}_w = 1 + \frac{M_o - M}{I_o} + \frac{M_o \theta}{I_o}\left\{1 + \frac{M}{M_o} - \frac{2}{\epsilon}\left[\frac{1 - (M/M_o)^{\epsilon}}{1 - M/M_o}\right]\right\} \tag{5}$$

(Figini, 8) where

$$\theta = K(1 - \rho)^2/(K + \rho)^2(\epsilon - 2)$$

$$\epsilon = 1 + (k_{qr}/k_q)(1 + K)^2/(K + \rho)I_o$$

$$\rho = k_r/k_q$$

when $\rho = 1$, $\bar{r}_w = 1 + \bar{r}_n$ as it must for the Poisson distribution. If the Q^* species cannot add monomer, that is, if $k_q = 0$, $\rho = \infty$, then

$$\theta = \frac{K}{\epsilon - 2} \qquad \epsilon = 1 + (k_{qr}/k_r)(1 + K)^2/I_o$$

If all the monomer has reacted, then

$$\bar{r}_n = 1 + M_o/I_o \tag{6}$$

$$\bar{r}_w = \bar{r}_n + M_o/I_o\ [\theta(\epsilon - 2)/\epsilon] \tag{7}$$

In order to have departures from the Poisson distribution, θ must be large and ϵ must be greater than 2. Similar equations have been derived by Szwarc and Hermans (10). If the transformation of R_n^* into Q_n^* is a second-order reaction

$$R_n^* + S \rightleftharpoons Q_n^* \tag{8}$$

and S is assumed to be invariant, k_{rq} can be replaced by $k'_{rq}S$ in the equations above.

Two propagating species exist in the stereospecific reactions:

$$\begin{aligned}
&\sim\sim A + M \xrightarrow{k_{aa}} \sim\sim A \\
&\sim\sim A + M \xrightarrow{k_{ab}} \sim\sim B \\
&\sim\sim B + M \xrightarrow{k_{bb}} \sim\sim B \\
&\sim\sim B + M \xrightarrow{k_{ba}} \sim\sim A
\end{aligned} \tag{9}$$

In this case, the assumption of an active initiator in equilibrium with an inactive initiator is unnecessary. Again

$$\bar{r}_n = 1 + (M_o - M)/I_o \tag{10}$$

but

$$\bar{r}_w = \frac{M_o - M}{I_o + M_o - M} + \frac{I_o + M_o - M}{I_o}$$

$$+ \alpha\left\{\frac{M_o - M}{I_o + M_o - M} - \frac{1 - \exp[-\beta(M_o - M)/I_o]}{\beta(I_o + M_o - M)/I_o}\right\} \tag{11}$$

(Figini, 9) where

$$\alpha = 2k_A r(q - 1)^2/k_{ab}(r + 1)^2(q + r)$$

$$\beta = k_{ab}(r + 1)^2/k_A(q + r)$$

$$r = k_{ba}/k_{ab}$$

$$q = (k_{bb} + k_{ba})/(k_{aa} + k_{ab})$$

$$k_A = k_{aa} + k_{ab}$$

The equation for $\bar{r}_w$ is quite similar to that presented in Section 3.9; for high molecular weights, $\bar{r}_w$ can be approximated by

$$\bar{r}_w \approx 1 + \bar{r}_n + \alpha(1 - [1 - \exp(-\beta\bar{r}_n)]/\bar{r}_n) \tag{12}$$

See figures given in Section 3.9.

5. Simultaneous Polymerization and Depolymerization

Rate of initiation equals $k_i MI$, monomer concentration may vary, no transfer, no termination.

Miyake and Stockmayer have considered the reactions

$$I + M \underset{k_i'}{\overset{k_i}{\rightleftharpoons}} M_1^* \tag{1}$$

$$M_n^* + M \underset{k_p'}{\overset{k_p}{\rightleftharpoons}} M_{n+1}^* \tag{2}$$

under a variety of conditions including continuous monomer feed and batch polymerization. We present here only a few of their results.

A. Monomer concentration invariant.

Setting $k_i = k_p$, $k_i' = k_p'$, we have

$$F(r) = \frac{\bar{r}_n}{\Phi_1(t)} \int_0^t \phi_r(t)\, dt \tag{3}$$

$$\bar{r}_n = \int_0^t \Phi_1(t)\, dt \Big/ \int_0^t \Phi_0(t)\, dt \tag{4}$$

$$\bar{r}_w = \int_0^t \Phi_2(t)\, dt \Big/ \int_0^t \Phi_1(t)\, dt \tag{5}$$

(Miyake and Stockmayer, 11) where

$$\phi_r(t) = t^{-1} \exp\{-(k_p' + k_p M_o)t\} \left[\frac{k_p M_o}{k_p'}\right]^{r/2} \left\{ r\tilde{I}_r(u) \right.$$

$$- \left[\frac{k_p M_o}{k_p'}\right]^{\frac{1}{2}} (r + 1)\, \tilde{I}_{r+1}(u) \Big\}$$

$$u = 2(k_p' k_p M_o)^{\frac{1}{2}} t$$

$\tilde{I}_r(u)$ is the modified Bessel function of the first kind of order $\underline{r}$.

$$\Phi_0(t) = \left[\frac{k_p M_o}{k_p'}\right]^{\frac{1}{2}} t^{-1} \exp\{-(k_p' + k_p M_o)t\}\tilde{I}_1(u)$$

$$\Phi_1(t) = k_p M_o - k_p' \int_0^t \Phi_0(t)\, dt$$

$$\Phi_2(t) = k_p M_o + k_p' \int_0^t \Phi_0(t)\, dt$$

$$- 2(k_p' - k_p M_o) \int_0^t \Phi_1(t)\, dt$$

When $k_p' > k_p M_o$, then after a long time the polymerization will cease at equilibrium. The distribution is

$$F(r)(\text{equilibrium}) = \left[1 - \frac{k_p M_o}{k_p'}\right]\left[\frac{k_p M_o}{k_p'}\right]^{r-1} \tag{6}$$

This is the Schulz-Flory distribution with $p = k_p M_o / k_p'$ and as usual

$$\bar{r}_n(\text{equilibrium}) = 1/(1 - p) \tag{7}$$

$$\bar{r}_w(\text{equilibrium}) = (1 + p)/(1 - p) \tag{8}$$

But, for $k_p' < k_p M_o$, equilibrium cannot be attained, and polymerization will continue to grow without limit, but $\bar{r}_w/\bar{r}_n$ will go through a maximum and will finally approach unity.

Some typical curves are given in Figure 3.5.1 for $\bar{r}_w/\bar{r}_n$ as a function of the reduced time variable $k_p M_o t$ for various values of $A = k_p'/k_p M_o$.

Curve	$k_p'/k_p M_o$	Limiting Value
a	1.2100	1.8260
b	1.0201	1.9800
c	1.0002	1.9998
d	0.8100	1.0000

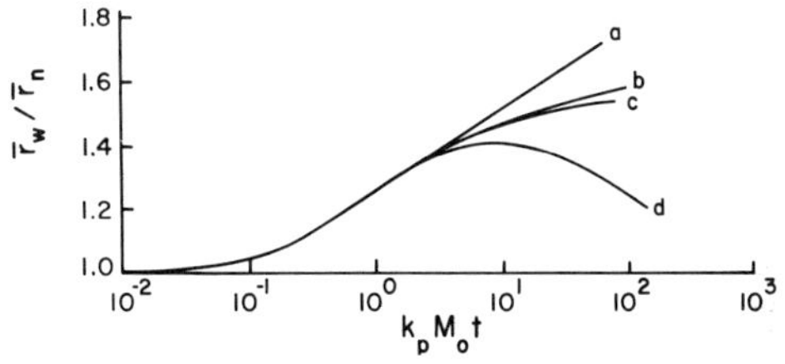

Figure 3.5.1 - The dispersion ratio $\bar{r}_w/\bar{r}_n$ as a function of the reduced time variable $k_p M_o t$ for simultaneous polymerization and depolymerization at constant monomer concentration. (Courtesy Makromol. Chem.) (Miyake and Stockmayer, 11).

The general case is quite complex, with $\phi_x(t)$ and $\Phi_0(t)$ taking different definitions depending on whether the term $k_i' - k_p' + (k_i + k_p)M_o$ is greater than, equal to, or less than $2(k_i k_p)^{\frac{1}{2}} M_o$. But again if $k_p' > k_p M_o$, the Schulz-Flory distribution results for exceedingly long times.

B. Monomer concentration varies.

When considering batch polymerizations, the authors had to resort to computer solutions of the differential equations. We present here one case, where $A = k_p'/k_p M_o = 0.1$ and $B = I_o/M_o = 0.1$.

This choice of values for A and B leads to small values of $\bar{r}_n$ and $\bar{r}_w$, 10.10 and 19.20 respectively, but they serve to show the general form of the solution. In Figure 3.5.2, $\bar{r}_n$, $\bar{r}_w$ and the ratio $\bar{r}_w/\bar{r}_n$ is plotted against the reduced time variable. The ratio reaches a maximum near 1.30, decreases to a minimum at 1.17, then finally increases to a equilibrium value of 1.90. In Figure 3.5.3, the fraction of unreacted initiator I/I_o passes through a minimum at 0.0061 and then increases to an equilibrium value of 0.10, whereas the residual monomer fraction M/M_o gradually decreases to 0.09. Miyake and Stockmayer present more detailed curves for various values of A while holding B = 0.1

The variation of I/I_o and M/M_o can be divided into three time periods: that of "Poisson distribution" when little material is consumed, an intermediate period when I/I_o approaches the minimum and M/M_o essentially reaches equilibrium, and finally the time required for I/I_o to approach its equilibrium value. With values of $k_p = 10^3$ liter/mole sec, $k_p' = 2 \times 10^{-4}$ sec^{-1}, $I_o = 2 \times 10^{-3}$ mole/liter, $M_o = 2$ mole/liter, reasonable for styrene at 25°, the first two time periods require only seconds, whereas attainment of near equilibrium would require almost a century.

$k_p'/k_p M_o = 0.1$ $\qquad$ $I_o/M_o = 0.1$

Curve	Variable	Limiting Value
a	$\bar{r}_n$	10.10
b	$\bar{r}_w$	19.20
c	$\bar{r}_w/\bar{r}_n$	1.90

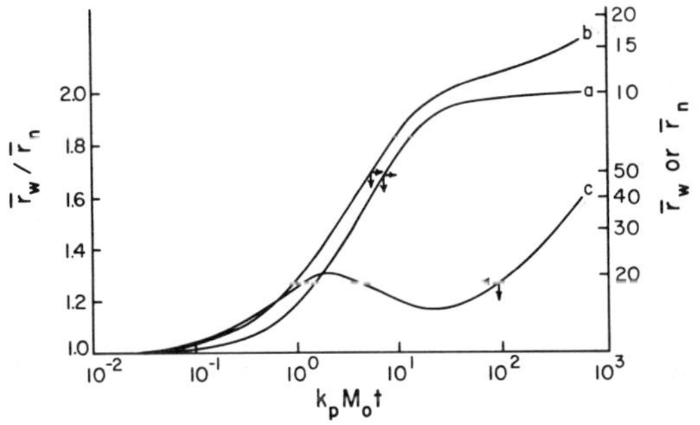

Figure 3.5.2 - The dispersion ratio, $\bar{r}_w/\bar{r}_n$ and the weight- and number-average degrees of polymerization as a function of the reduced time variable k_pM_ot for simultaneous polymerization and depolymerization with varying monomer concentration. (Courtesy Makromol. Chem.) (Miyake and Stockmayer, 11).

Curve	Variable	Limiting Value
a	M/M_o	0.09
b	I/I_o	0.10

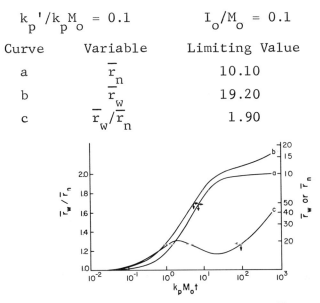

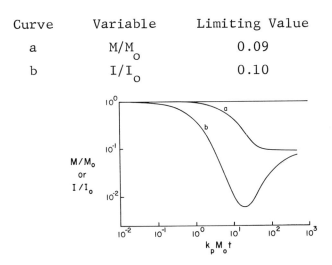

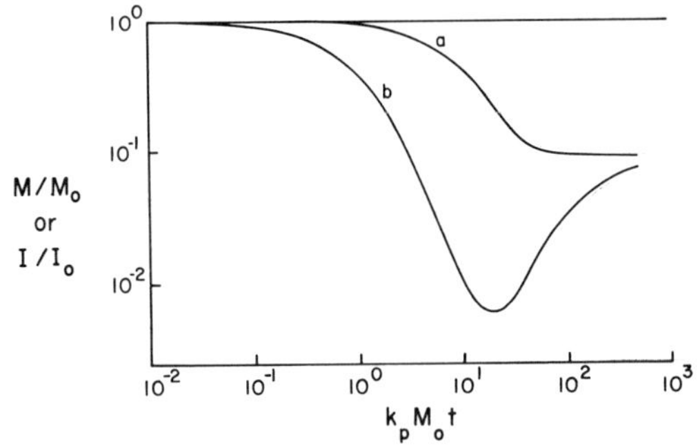

Figure 3.5.3 - Fractional initiator concentration and fractional monomer concentration as a function of the reduced time variable k_pM_ot. Same conditions as in Figure 3.5.2. (Courtesy *Makromol. Chem.*) (Miyake and Stockmayer, 11).

6. The k_p Varies with Chain Length

Rate of initiation equals k_iMI, monomer concentration varies, no transfer, no termination.

We consider now the distribution which results when the propagation rate constant depends on molecular weight. (See also Section 4.3.) To avoid confusion, we define k_o as the rate constant for initiation, and k_1, k_2, k_3, and so on, as the rate constants for chains containing 1, 2, 3, and so on, monomer units. The concentration of polymer molecules of size r is given by

$$P_r = (k_{r-1}/k_o)(I/I_o)^{k_r/k_o} \int_{I/I_o}^{1} P_{r-1} \times (I/I_o)^{-[(k_r/k_o)+1]} \, d(I/I_o) \qquad (1)$$

(Maget, 12).

If all k's are equal to k_o, this equation reduces to the Poisson distribution.

If all the propagation constants, k_r, are different, then

$$P_r = I_o(-1)^r \prod_{s=0}^{r-1} \left(\frac{k_s}{k_o}\right) \sum_{s=0}^{r} \frac{(I/I_o)^{k_s/k_o}}{\prod_{\substack{q=0\\q\neq s}}^{r} (k_s - k_q)/k_o} \quad (2)$$

(Maget, 12). This function is useful for a small number of reaction steps such as the total chlorination of an alkane.

If all k's vary according to the relation

$$k_0:k_1:k_2:k_3: \ldots k_n = m:(m-1):(m-2):(m-3):\ldots 1 \quad (3)$$

in which there are only a finite number of reaction steps, n, then the distribution function becomes

$$P_r = I_o(I/I_o)(m/r)[(I_o/I)^{1/m} - 1]^r \quad (4)$$

(Maget, 12). If

$$k_1 \neq k_2 \neq k_3 \ldots \neq k_m = k_{m+1} = k_{m+2} = \ldots = k_n \quad (5)$$

that is, the low molecular weight species have rate constants which depend on molecular size, but the large molecules have attained an asymptotic value k_n. The distribution for $r < m$ is given by equation (2); for $r \geqslant m$

$$P_r = I_o(-1)^r \left[\frac{k_m}{k_o}\right]^{r-m} \left[\frac{I}{I_o}\right]^{k_m/k_o} \prod_{s=0}^{m-1} \left[\frac{k_s}{k_o}\right]$$

$$\times \sum_{s=0}^{m-1} \sum_{q=r-m+1}^{\infty} \frac{\{[(k_s - k_m)/k_o] \ln (I/I_o)\}^q/q!}{\left[\dfrac{k_s - k_m}{k_o}\right]^{r-m} \sum\limits_{\substack{t=0 \\ t \neq s}}^{m} \left[\dfrac{k_s - k_t}{k_o}\right]} \tag{6}$$

(Maget, 12). This equation reduces to that in Section 1 when $k_o = k_p$ and to that in Section 2 when $k_o \neq k_p$, k_p does not vary with chain length.

If the first three reaction steps have different reactivities, that is, $k_o \neq k_1 \neq k_2 = k_m$, then

$$P_r = \frac{I_o(-1)^r \, (k_m/k_o)^{r-2} k_1}{(k_o - k_1)}$$

$$\times \left[\frac{\dfrac{I}{I_o} - \left[\dfrac{I}{I_o}\right]^{k_m/k_o} \sum\limits_{s=0}^{r-2} [(1 - k_m/k_o) \ln (I/I_o)]^s/s!}{(1 - k_m/k_o)^{r-1}} \right.$$

$$\left. - \frac{\left[\dfrac{I}{I_o}\right]^{k_1/k_o} - \left[\dfrac{I}{I_o}\right] \sum\limits_{s=0}^{r-2} \{[(k_1 - k_m)/k_o] \ln (I/I_o)\}^s/s!}{[(k_1 - k_m)/k_o]^{r-1}} \right] \tag{7}$$

(Maget, 12). In view of the complexity of the Gold distribution, no attempt is made to present graphs of this function.

7. Deactivation by Transfer to Monomer

Rate of initiation is instantaneous or equals $k_i MI$, monomer concentration varies, transfer to monomer, no termination (5, 13-18).

Another means of deactivating the chain and thereby broadening the distribution is by invoking the transfer-to-monomer reaction

$$R_r^* + M \rightarrow P_r + R_o^* \qquad k_{tr,m} \tag{1}$$

in addition to the usual nonterminating reactions

$$I^* + M \rightarrow R_1^* \qquad k_i \tag{2}$$

$$R_r^* + M \rightarrow R_{r+1}^* \qquad k_p \tag{3}$$

In the counting scheme used here, initiator fragments are included in the number of repeat units, hence R_1^* contains two units, either a monomer and an initiator fragment or two monomer units as a result of transfer. The differential equation can be linearized by letting τ be defined by

$$t = \int_0^{\tau} \frac{ds}{(k_p + k_{tr,m})M(s)} \tag{4}$$

With the following definitions

$$\mu = k_i/(k_p + k_{tr,m})$$

$$\sigma = k_{tr,m}/(k_p + k_{tr,m}) \tag{5}$$

$$\lambda = 1 - \sigma$$

and the constraint that the total number of active molecules must equal the number of initiator molecules that have disappeared

$$I_o - I(\tau) = \sum_{r=0}^{\infty} R_r^* \tag{6}$$

The concentration of active monomer formed by transfer, R_o^*, and unreacted monomer are

$$I(\tau) = I_o \exp(-\mu\tau) \tag{7}$$

$$R_o^*(\tau) = \sigma I_o \{1 - [e^{-\mu\tau}/(1 - \mu)] + [\mu/(1 - \mu)]e^{-\tau}\} \tag{8}$$

$$M(\tau) = M_o + I_o\left\{\frac{(1 - \mu)^2 - \sigma^2}{\mu(1 - \mu)}(1 - e^{-\mu\tau}) + [\mu\sigma^2/(1 - \mu)](1 - e^{-\tau}) + (\sigma^2 - 1)\tau\right\} \tag{9}$$

and the concentration of active molecules is

$$R_r^* = \lambda^{r-1} I_o \left\{ A \int_0^{\tau} \frac{e^{-s} s^{r-1}}{(r-1)!} ds \right.$$

$$\left. + \frac{Be^{-\mu\tau}}{(1-\mu)^r} \int_0^{(1-\mu)\tau} \frac{e^{-s} s^{r-1}}{(r-1)!} ds + D\frac{\tau^r e^{-\tau}}{r!} \right\} r \geqslant 1 \qquad (10)$$

(Kyner, Radok, and Wales, 13) where

$$A = \sigma\lambda$$

$$B = \mu - [\sigma\lambda/(1-\mu)] \qquad (11)$$

$$D = \mu\sigma\lambda/(1-\mu)$$

and the integral is the incomplete gamma function, whereas the concentration of inactive molecules is

$$P_r = \sigma \int_0^{\tau} R_r^* \, d\tau \qquad (12)$$

The number and weight-average degrees of polymerization are given by

$$\bar{r}_n = [\lambda\mu(1+\sigma)(1-\mu)\tau + (1+\sigma-2\mu)(\mu-\lambda)(1-e^{-\mu\tau})$$

$$+ \mu^2(1-e^{-\tau})]/[\lambda\mu\sigma(1-\mu)\tau + (\sigma-\mu)(\mu-\lambda)(1-e^{-\mu\tau})$$

$$+ \lambda\mu^2\sigma(1-e^{-\tau})] \qquad (13)$$

$$\bar{r}_w = 1 + \Big\{2\lambda(a_1 + a_2 + a_3 + \sigma)\tau + 2\{1 - (\lambda a_1/\mu)$$

$$- [\lambda\sigma/\mu(1 - \mu)]\}(1 - e^{-\mu\tau}) - (2\lambda a_2/\sigma)(1 - e^{-\sigma\tau})\Big\}$$

$$\div \Big\{\{[(1 + \sigma - 2\mu)(\mu - \lambda)]/\mu(1 - \mu)\}(1 - e^{-\mu\tau})$$

$$+ \lambda(1 + \sigma)\tau + [\lambda^2\mu\sigma/(1 - \mu)](1 - e^{-\tau})\Big\} \tag{14}$$

$$a_1 = \{2\mu - [\lambda\sigma/(1 - \mu)] - \lambda\}/(\mu - \sigma) \qquad a_3 = \sigma\mu/(1 - \mu)$$

$$a_1 + a_2 + a_3 = (\lambda^2 + 2\lambda\sigma)/\sigma$$

By use of a simpler kinetic scheme, Nanda (14) has derived somewhat simpler expressions which are valid for long chains. His equations, however, are still quite complex.

All of these equations reduce to the Gold distribution, Section 2, when $\sigma = 0$. Inspection of equations (13) and (14) shows that initially, at small values of τ, a Poisson distribution results, but as conversion increases the distribution continually broadens. If at infinite time, τ is also infinite, all polymerization has ceased and the final distribution is the Schulz-Flory distribution. However, inspection of the definition of τ, equation (4), shows that τ may still be finite when $t = \infty$. Chiang and Hermans (5) have shown how to obtain the moments of the distribution, ΣP_r, ΣR_r^*, ΣrP_r, ΣrR_r^*, Σr^2P_r, $\Sigma r^2R_r^*$, and so on, directly from the kinetic scheme without a knowledge of the individual concentrations and without having to sum these equations. If we assume instantaneous initiation, then the number- and weight-average degrees of polymerization are

$$\bar{r}_n = (1 + \tau)/(1 + \sigma\tau) \tag{15}$$

$$\bar{r}_w = \{1 + \tau(1 + \lambda)/\sigma - 2(\lambda/\sigma)^2(1 - e^{-\sigma\tau})\}/(1 + \tau) \tag{16}$$

At infinite time, all monomer is consumed but living radicals still exist, $\tau = M_o/I_o$ [see equation (23) below]. Figure 3.7.1 shows the variation of $\bar{r}_n/\bar{r}_w$ with σ for $\bar{r}_n = 100$. The ratio varies from 1.01 for no transfer to 2.00 when transfer predominates.

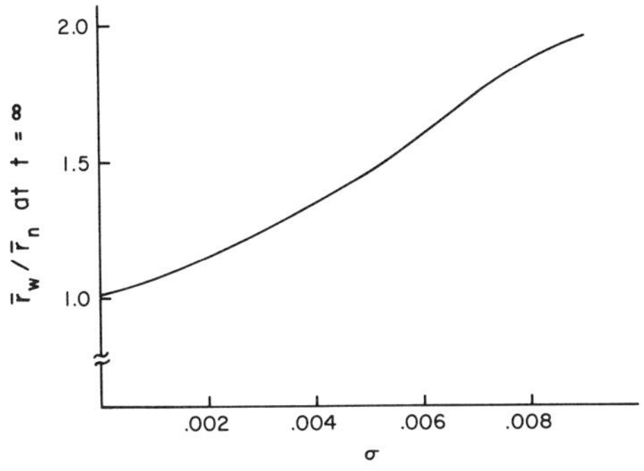

Figure 3.7.1 - Dispersion ratio $\bar{r}_w/\bar{r}_n$ as a function of σ with $\bar{r}_n = 100$, for instantaneous initiation, transfer to monomer, no termination, at infinite time, that is, all monomer is consumed. (After Chiang and Hermans, 5).

When instananeous initiation is assumed, the differential kinetic equations becomes easier to solve. Here the counting scheme is altered so that R_1^* may or may not contain an initiator fragment, depending on the mode of formation. The resulting equations for active and inactive polymer molecules are

$$R_1^* = I_o\{\sigma + \lambda e^{-\tau}\} \tag{17}$$

$$R_2^* = \lambda I_o\{\sigma(1 - e^{-\tau}) + \lambda\tau e^{-\tau}\} \tag{18}$$

$$R_r^* = \lambda^{r-1} I_o\{\sigma(1 - e^{-\tau}) + \frac{\lambda\tau^{r-1} e^{-\tau}}{(r - 1)!}$$

$$-\sigma \sum_{s=1}^{r-2} \tau^s e^{-\tau}/s!\}, \qquad r \geqslant 3 \tag{19}$$

$$P_1 = \sigma I_o\{\sigma\tau - \lambda(e^{-\tau} - 1)\} \tag{20}$$

$$P_2 = \sigma\lambda I_o\{\sigma(\tau + e^{-\tau}) + \lambda P(2\tau \mid 4)\} \tag{21}$$

$$P_r = \sigma\lambda^{r-1} I_o\{\sigma(\tau - 1 + e^{-\tau}) + \lambda P(2\tau \mid 2r)$$

$$- \sigma \sum_{s=2}^{r-1} P(2\tau \mid 2s)\}, \qquad r \geqslant 3 \tag{22}$$

(Peebles, 16) where τ is now the integrated form of equation (4)

$$\tau = (M_o/I_o)[1 - \exp\{-(k_p + k_{tr,m})I_o t\}] \tag{23}$$

and

$$P(2\tau \mid 2r) = \int_0^\tau \frac{\tau^{r-1} e^{-\tau}}{(r - 1)!} d\tau \tag{24}$$

where $P(\chi^2 \mid \nu)$ is the chi-square probability function. Some of the properties of $P(2\tau \mid 2r)$ can be used to obtain the approximations

$$R_r^* \approx \lambda^{r-1} I_o\{\sigma P(x) + \frac{\tau^{r-1} e^{-\tau}}{(r - 1)!}\} \tag{25}$$

$$P_r \approx \sigma\lambda^{r-1} I_o\{\sigma(\tau + 1 - r) + \lambda\} \qquad r \leqslant r_j \tag{26}$$

$$\approx \sigma\lambda^{r-1} I_o\{\sigma(\tau + 1 - r_j) + \lambda P(x)$$

$$- \sigma \sum_{s=r_j}^{r-1} P(x_s)\} \qquad r > r_j \tag{27}$$

(Peebles, 16) where $P(x)$ is the normal or Gaussian probability function

$$P(x) = (2\pi)^{-\frac{1}{2}} \int_{-\infty}^{x} e^{-t^2/2} dt \tag{28}$$

$$x = (\tau - r)/r^{\frac{1}{2}} \tag{29}$$

$$x_s = (\tau - s)/s^{\frac{1}{2}} \tag{30}$$

and r_j is selected so that P(x) can be replaced by unity when $r < r_j$, that is, when $x \geqslant 3.1$, $1 > P(x) \geqslant 0.99903$. Equations (25) and (27) are plotted in Figure 3.7.2 for $\bar{r}_n = 100$, $\sigma = 0.005$ to show how the relative concentration of R_r^* and P_r vary with r. The weight-fraction distributions corresponding to equations (25) to (27) are found by use of the Chiang and Hermans procedure; thus

$$\Sigma r P_r + \Sigma r R_r^* = I_o(1 + \tau) \tag{31}$$

Figure 3.7.3 shows the weight distribution for active and inactive polymer with the conditions $\bar{r}_n = 100$, $\sigma = 0$, 0.001, and 0.005. Curves b and d are almost coincidental up to r = 70. The weight distributions for inactive polymer is shown in Figure 3.7.4 on an expanded scale.

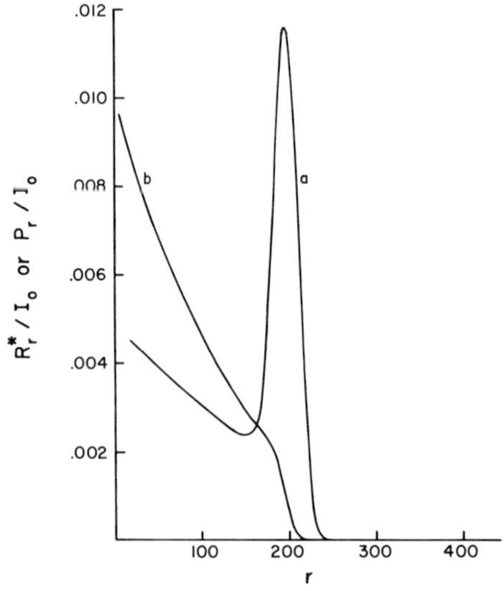

Figure 3.7.2 - Variation of relative concentration of active (a) and inactive (b) species as a function of r for

instantaneous initiation, transfer to monomer, and no termination. $\lambda = 0.995$, $\bar{r}_n = 100$ (after Peebles, 16).

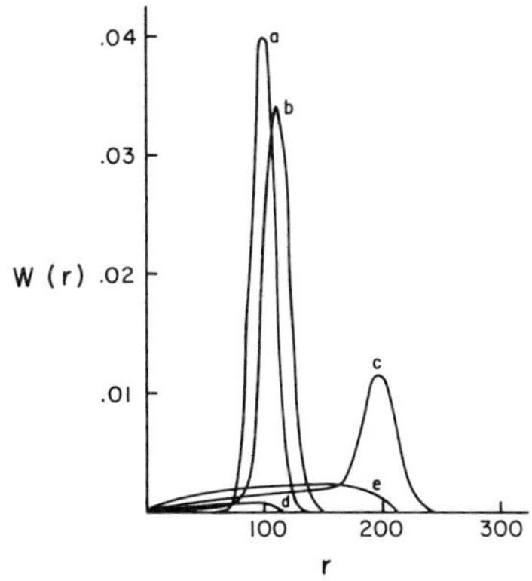

Figure 3.7.3 - The weight fraction of active and inactive chains as a function of r for instantaneous initiation, transfer to monomer, no termination, and $\bar{r}_n = 100$. Curves a, b, c give $rR_r^*/\Sigma(rP_r + rR_r^*)$ for $\lambda = 1.000$, 0.999, and 0.995 respectively. Curves d and e give $rP_r/\Sigma(rP_r + rR_r^*)$ for $\lambda = 0.999$ and 0.995, respectively. Curve a is the Poisson distribution when transfer does not occur. (Courtesy J. Polymer Sci.) (Peebles, 16).

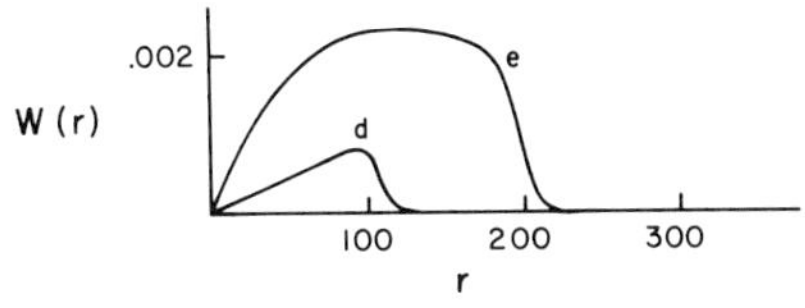

Figure 3.7.4 - The weight fraction of inactive polymer as a function of r on an expanded scale. Same conditions as in Figure 3.7.3 (after Peebles, 16).

If initiation forms a propagating species with two active centers, R_r^{**}, transfer to monomer occurs to create a species with one active center $S_1{}^*$ and results in a monoactive, $S_r{}^*$, or an inactive chain, P_r, and termination does not occur, then the distribution equations are even more complex than for the situation already considered. The results are given by Nanda and Jain (17). However, if instantaneous initiation is again assumed, the results are much simpler. Using the procedures outlined earlier, we find

$$M = M_o \exp\{-(k_p + k_{tr,m})I_o t\}$$

$$= M_o - I_o \tau \qquad (32)$$

provided that the act of transfer between an $S_1{}^*$ center and a monomer unit produces another $S_1{}^*$ center and a nonpolymerizeable species P_1. If we assume that P_1 has the same reactivity as monomer, then slightly more complex equations are obtained.

$$R_r^{**} = (\tfrac{1}{2}I_o)(2\lambda\tau)^{r-2}e^{-2\tau}/(r-2)! \qquad r \geqslant 2 \qquad (33)$$

$$S_1{}^* = I_o\sigma(1 - e^{-\tau}) \qquad (34)$$

$$S_r{}^* = \sigma I_o \lambda^{r-2}\{\lambda + \sigma \sum_{s=0}^{r-1} \tau^s e^{-\tau}/s! - \tau^{r-1}e^{-\tau}/(r-1)!$$

$$- \sum_{s=0}^{r-2} (2\tau)^s e^{-2\tau}/s!\} \qquad r \geq 2 \tag{35}$$

$$\approx \sigma I_o \lambda^{r-2}\{\lambda - \tau^{r-1} e^{-\tau}/(r - 1)!\} \qquad r \leq r_j \tag{36}$$

$$\approx \sigma I_o \lambda^{r-2}\{P[2\tau - r + 1)/(r - 1)^{\frac{1}{2}}]$$

$$- \sigma P[(\tau - r)/r^{\frac{1}{2}}] - \tau^{r-1} e^{-\tau}/(r - 1)!\} \qquad r > r_j$$

$$\tag{37}$$

$$P_1 = \sigma^2 I_o(\tau + e^{-\tau} - 1) \tag{38}$$

$$P_r = \sigma^2 I_o \lambda^{r-2}\{\lambda\tau + \sigma \sum_{s=1}^{r} P(2\tau \mid 2s) - P(2\tau \mid 2r)$$

$$- \tfrac{1}{2} \sum_{s=1}^{r-1} P(4\tau \mid 2s)\} \qquad r \geq 2 \tag{39}$$

$$\approx \sigma^2 I_o \lambda^{r-2}\{\lambda\tau - (\tfrac{1}{2} - \sigma)r + \tfrac{1}{2}\} \qquad r \leq r_j \tag{40}$$

$$\approx \sigma^2 I_o \lambda^{r-2}\{\lambda\tau - (\tfrac{1}{2} - \sigma)r_j + \sigma \sum_{s=r_j+1}^{r} P[(\tau - s)/s^{\frac{1}{2}}]$$

$$- P[(\tau - r)/r^{\frac{1}{2}}] \qquad - \tfrac{1}{2} \sum_{s=r_j+1}^{r-1} P[(2\tau - s)/s^{\frac{1}{2}}]\}$$

$$r > r_j \tag{41}$$

(Peebles, 18).

$$\bar{r}_n = \frac{\Sigma(rR_r^{**} + rS_r^{*} + rP_r)}{\Sigma(R_r^{**} + S_r^{*} + P_r)} = \frac{(1 + \tau)}{(\frac{1}{2} + \sigma\tau)} \tag{42}$$

$$\bar{r}_w = (1/\sigma^2)\{4\sigma - 1 - \sigma^2 + (2\sigma - \sigma^2)\tau$$

$$+ (2\sigma^2 - 2\sigma)\theta + (1 - 2\sigma + \sigma^2)\theta^2\}/(1 + \tau) \tag{43}$$

where

$$\theta = \exp(-\sigma\tau)$$

and τ is given by equation (23), r_j is selected so that $P(x)$ can be replaced by unity when $r \leqslant r_j$, and $P(x)$ is given by equation (28).

The average degrees of polymerization and the weight fraction distribution for any combination of species can be calculated with the aid of the following equations:

$$\Sigma R_r^{**} = \tfrac{1}{2}I_o\theta^2; \qquad \Sigma S_r^{*} = I_o(1 - \theta^2)$$

$$\Sigma P_r = I_o\{\sigma\tau + (\theta^2 - 1)/2\}$$

$$\Sigma rR_r^{**} = I_o(1 + \lambda\tau)\theta^2$$

$$\Sigma rS_r^{*} = I_o\{(1/\sigma) + \theta - (1 + (1/\sigma) + 2\lambda\tau)\theta^2\} \tag{44}$$

$$\Sigma r P_r = I_o\{1 - (1/\sigma) - \theta + \theta^2/\sigma + \lambda\tau\theta^2 + \tau\}$$

$$\Sigma r^2 R_r^{**} = I_o(5\lambda\tau + 2\lambda^2\tau^2 + 2)\theta^2$$

$$\Sigma r^2 S_r^{*} = (I_o/\sigma^2)\{1 + \lambda + \sigma(2 + \sigma)\theta + 2\sigma^2\lambda\tau\theta$$

$$- (2 + \sigma + \sigma^2)\theta^2 - 2\lambda\sigma(2 + 3\sigma)\tau\theta^2 - 4\lambda^2\sigma^2\tau^2\theta^2\}$$

$$\Sigma r^2 P_r = (I_o/\sigma^2)\{(5\sigma - 3 - \sigma^2) + \sigma(2 - \sigma)\tau$$

$$- \sigma(4 - \sigma)\theta - 2\lambda\sigma^2\tau\theta + (3 - \sigma)\theta^2$$

$$+ (4\sigma - 3\sigma^2 - \sigma^3)\tau\theta^2 + 2\lambda^2\sigma^2\tau^2\theta^2\}$$

Figures 3.7.5 and 3.7.6 present the weight fraction distributions for diactive, monoactive, and inactive polymer for $\bar{r}_n = 100$, $\sigma = 0$, 0.001, and 0.005. The total weight fraction of each species is given in Figure 3.7.6.

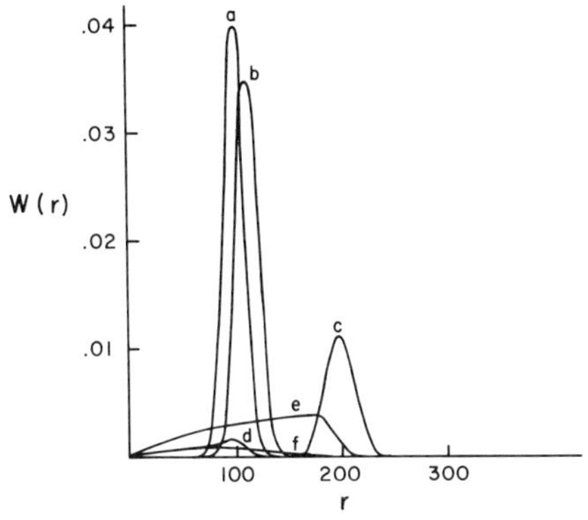

Figure 3.7.5 - The weight fraction of diactive, monoactive and inactive chains as a function of r for instantaneous initiation to form chains with two active ends, transfer to monomer which forms chains with either one or no active end and no termination for $\bar{r}_n = 100$. Curves a, b, and c represent the weight fraction of diactive chains for $\lambda = 1.000$, 0.999, and 0.995, respectively. Curves d and e represent the monoactive chains for $\lambda = 0.999$ and 0.995. Curve f represents the inactive chains for $\lambda = 0.995$. (Courtesy J. Polymer Sci.) (Peebles, 18).

Weight Fraction of each Molecular Type for $\bar{r}_n = 100$.

σ	$\bar{r}_w$	τ	R_r^{**}	S_r^{*}	P_r
0	101.0	49.00	1.000	--	--
.001	108.2	54.44	.896	.102	.002
.003	124.6	70.00	.655	.321	.024
.005	146.7	98.00	.373	.525	.102

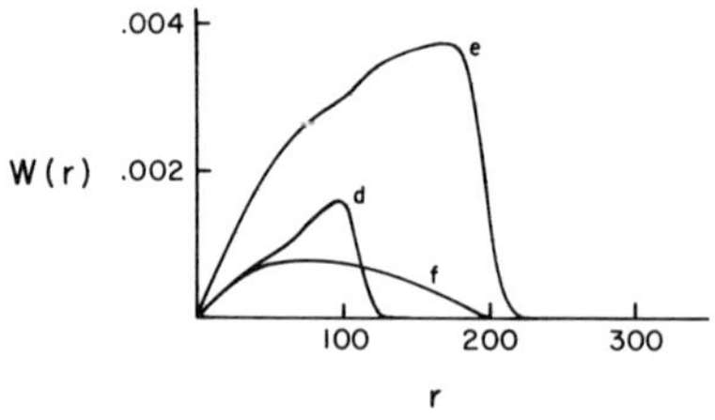

Figure 3.7.6 - The weight fraction of monoactive and inactive polymer as a function of r on an expanded scale. Same conditions as in Figure 3.7.5. (after Peebles, 18)

8. Deactivation by Transfer to Monomer and Slow First-Order Termination

Instantaneous initiation, monomer concentration varies, transfer to monomer, termination by a slow first-order reaction.

Chiang and Hermans have considered this case in which dead polymer is formed by two mechanisms: transfer to monomer which does not deactivate propagation, and a slow first-order deactivation. The total active polymer concentration is

$$R^* = I_o \exp(-k_t t) \tag{1}$$

At infinite time, no active chains remain; the final monomer concentration is

$$M_f = M_o \exp\{-(k_p + k_{tr,m})I_o/k_t\} \tag{2}$$

$$\bar{r}_n(t = \infty) = \left\{1 + (M_o/I_o)[1 - \exp\{-(k_p + k_{tr,m})I_o/k_t\}]\right\}$$
$$\div \left\{1 + [k_{tr,m}/(k_p + k_{tr,m})](M_o/I_o)\right.$$
$$\left.\times [1 - \exp\{-(k_p + k_{tr,m})I_o/k_t\}]\right\} \tag{3}$$

$$\bar{r}_w\ (t = \infty) = [I_o + (1 + 2k_p/k_{tr,m})(M_o - M_f)$$
$$- 2(k_p^2/k_t k_{tr,m})M_o \exp\{-(k_p + k_{tr,m})I_o/k_t\}$$

$$\times \int_{-\infty}^{0} R^{*} \exp(\lambda \Phi)\, d\Phi]/[I_o + M_o] \quad (4)$$

(Chiang and Hermans, 5) where

$$\lambda = k_{tr,m} M_f / k_t$$

$$\Phi(I) = Ei[(k_p + k_{tr,m})R^{*}/k_t] - Ei[(k_p + k_{tr,m})I_o/k_t]$$

$$Ei(x) = \int_{-\infty}^{x} (e^y/y)\, dy$$

To find the value of the integral in $\bar{r}_w$, first plot R^* versus Φ, then plot $R^* \exp(\lambda\Phi)$ versus Φ, and find the area under the curve.

9. Deactivation by Initiator Expulsion Reaction

Instantaneous initiation, monomer concentration invariant, no transfer, no termination.

We assume here that the growing radicals can interact with the initiator in the system to produce dead polymer and active initiator fragments. The interaction is usuall written as a transfer-to-initiator reaction, but the mecha nism of the reaction is not similar to the transfer-to-initiator reaction which occurs in free-radical systems. To distinguish these cases, we call the present case the initiator expulsion reaction. The number of active chains is always equal to the initial charge of initiator. The distribution functions have not been solved, but the moments of the distribution have been found:

$$\bar{r}_n = \frac{(k_pM + B)t}{Bt + [k_pM/(k_pM + B)][1 - \exp\{-(k_pM + B)t\}]} \qquad (1)$$

$$\approx k_pMt/(1 + Bt) \qquad (2)$$

a short time after polymerization has started, where

$$B = k_{iex} \qquad (3)$$

$$= k_{tr,i}\, I + k_{tr,a}A^{\frac{1}{2}} \qquad (4)$$

in the case considered by Chien (19).

$$\bar{r}_w = 1 + (2k_pM/B)[1 - (1 - \exp\{-Bt\})/Bt] \qquad (5)$$

At infinite time

$$\bar{r}_n\ (t{=}\infty) = k_pM/B \qquad (6)$$

$$\bar{r}_w\ (t{=}\infty) = 1 + 2\,\bar{r}_n\,(t{=}\infty) \qquad (7)$$

In Figure 3.9.1 $\bar{r}_n$ is plotted against time for a variety of $\underline{B}$ values. To ensure that equation (2) is valid, we arbitrarily required $\bar{r}_n$ to equal 100 at 25 units of time.

Figure 3.9.2 presents the variation of $\overline{r}_w/\overline{r}_n$ with time.

Curve	A	B	Limiting Value
a	4	0	∞
b	4.01	.001	40,100
c	4.1	.001	4,100
d	5	.01	500
e	14	.1	140

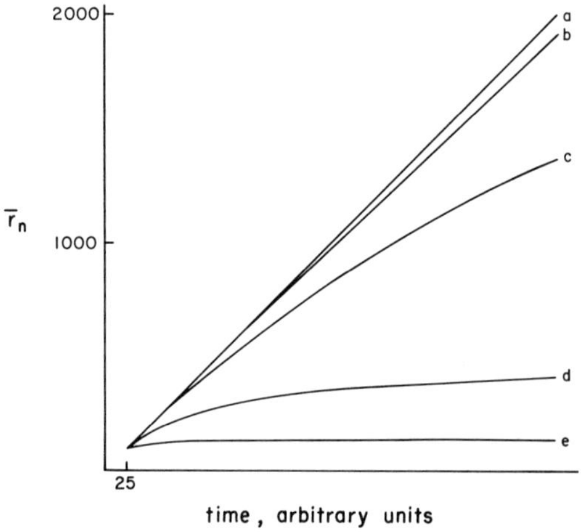

Figure 3.9.1 - $\overline{r}_n$ as a function of time, in arbitrary units, for various values of $A = k_pM$, $B = k_{iex}$, instantaneous initiation, no termination, constant monomer concentration. (After Chien, 19).

10. References

1. P. J. Flory, "Molecular Size Distribution in Ethylene Oxide Polymers," J. Am. Chem. Soc., 62, 1561 (1940).
2. L. Gold, "Statistics of Polymer Molecule Size Distributions for an Invariant Number of Propagating Chains," J. Chem. Phys., 28, 91 (1958).

Curve	B
a	0
b	.0001
c	.001
d	.01
e	.1

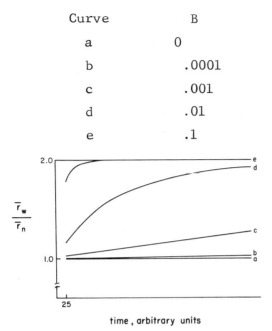

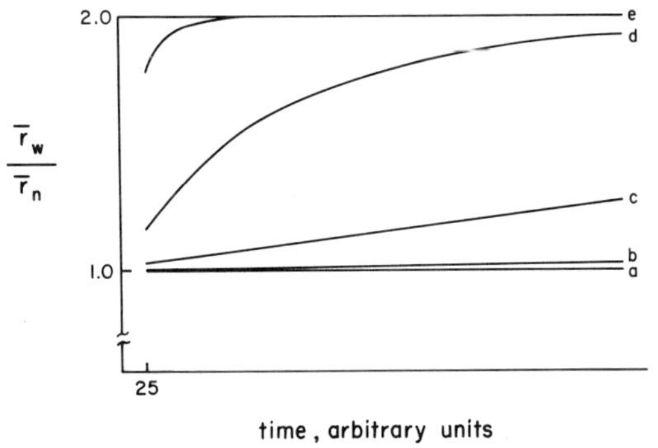

Figure 3.9.2 - Dispersion ratio as a function of time, in arbitrary units. $k_p M$ and $\underline{B}$ adjusted to make $\bar{r}_n = 100$ at $t = 25$. Same conditions as in Figure 3.9.1.

3. B. D. Coleman, F. Gornick, and G. Weiss, "Statistics of Irreversible Termination in Homogeneous Anionic Polymerization," J. Chem. Phys., 39, 3233 (1963).

4. T. A. Orofino and F. Wenger, "The Effect of Impurities on the Molecular Weight Distributions of Anionic Polymers," J. Chem. Phys., 35, 532 (1961).

5. R. Chiang and J. J. Hermans, "Influence of Catalyst Depletion or Deactivation on Polymerization Kinetics. II. Nonsteady-State Polymerization," J. Polymer Sci., A1, 4, 2843 (1966).

6. A. Guyot, "Etude Theorique de la Distribution des Masses Moleculaires dans un Cas de Polymerisation sans Regime Stationnaire," J. Chim. Phys., 1964, 548.
7. See Ref. 8, Chapter 2.
8. R. V. Figini, "Statistische Berechnung über den Wachstumsprozess von Polymerketten mit Wechselnder Aktivität," Makromol. Chem., 71 , 193 (1964).
9. R. V. Figini, "Molekulargewichtsverteilungen bei Anionischer Polymerisation nach dem Einwegmechanismus unter Berücksichtigung Stereospezifischer Anlagerung des Monomeren," Makromol. Chem., 88, 272 (1965).
10. M. Szwarc and J. J. Hermans, "Molecular Weight Distribution in a Non-terminated Polymerization Involving Living and Dormant Polymers," J. Polymer Sci., B2, 815 (1964).
11. A. Miyake and W. H. Stockmayer, "Theoretical Reaction Kinetics of Reversible Living Polymerization," Makromol. Chem., 88, 90 (1965).
12. H. J. R. Maget, "Product Distribution of Consecutive Competitive Second-Order Reactions," J. Polymer Sci., A2, 1281 (1964).
13. W. T. Kyner, J. R. M. Radok, and M. Wales, "Kinetics and Molecular Weight Distributions for Unsteady-State Polymerizations Involving Termination by Chain Transfer with the Monomer," J. Chem. Phys., 30, 363 (1959).
14. V. S. Nanda, "Theoretical Study of the Effect of Initiation and Transfer Rates on Size Distribution in Anionic Polymers," Trans. Faraday Soc., 60, 949 (1964).

15. A. Guyot, "Calcul Theorique de la Distribution des Masses Moleculaires d'une Polymerisation avec Amorcage Instante, Transfert sur le Monomere et Pas D'arret," Polymer Letters, 6, 123 (1968).

16. L. H. Peebles, Jr., "The Molecular Weight Distribution for Polymers Formed by Instantaneous Initiation, Transfer to Monomer, and No Termination," Polymer Letters, 7, 75 (1969). (See ref. 18.)

17. V. S. Nanda and R. K. Jain, "Effect of Initiation and Transfer Rates on Molecular Size Distribution in Dianionic Polymers," Trans. Faraday Soc., 64, 1022 (1968).

18. L. H. Peebles, Jr., "The Molecular Weight Distribution for Diactive Polymers Formed by Instantaneous Initiation, Transfer to Monomer, and No Termination," J. Polymer Sci., A2, 8, 1235 (1970). (Note: equation 13 of reference 16 should be written $x = (\tau - r)/r^{\frac{1}{2}}$; equations there and in this paper containing x should be corrected. The figures are correct.)

19. J. C. W. Chien, "Kinetics of Propylene Polymerization Catalyzed by α-Titanium Trichloride-Diethylaluminum Chloride," J. Polymer Sci., A1, 425 (1963).

Chapter 4

Linear Condensation Polymerization Without Ring Formation

Contents

1. Introduction

Condensation polymers are usually formed by reacting

two molecules together to produce a single larger molecule and a very small by-product molecule. Examples are reactions of carboxylic acids and hydroxyl groups to form a polyester and of carboxylic acids and amines to form a polyamide. In each of these cases, water is the by-product. Condensation-type polymers can also be formed by ring opening polymerizations such as a lactam or anhydride reacting with an active group. In the latter case, a by-product is not formed.

The simplest case, that of an AB type molecule, such as an ω-hydroxy acid, was considered by Flory in 1936 (1); the distribution is described in Chapter 1, Section 2

In this chapter, we denote functional groups by the capital letters A, B, C, and so on, and the extent of reaction of these groups by lower case Greek letters α, β, γ, and so on, where $\alpha = 1 - (A/A_o)$, $\beta = 1 - (B/B_o)$, and so on. A prime, that is, β', γ', means that some type of reaction must precede the reaction under consideration. Thus if an amine initiator is used in lactam polymerization opening of the lactam ring must occur before the newly created amine can react. The symbol AA represents a monomer with two functional groups of type A, no distinction being made as to the number of links connecting the two A's in the monomer.

Because the weight-average molecular weight will depend upon the molecular weights of the original monomers used and the molecular weights of the by-products, the symbol W_{AA} represents the molecular weight of the reacted monomer molecule AA. The function W(r) can be considered exact if the weight of the end groups can be neglected relative to the polymer. If W(r) is desired for very low molecular weight polymers or oligomers, a better procedure is to calculate it directly from F(r) than to use the given formulas.

2. Simple Linear Condensation. AA reacting with BB. The nylon case of hexamethylenediamine and adipic acid.

Polymers of this form are denoted by Flory as "type II." The distribution was published in 1936 (1). The following formulas were taken from Case (2).

Let α = fraction of A groups that have reacted

β = fraction of B groups that have reacted

The extent of reaction depends on the initial quantity of both monomers present, hence

$$\alpha(AA)_o = \beta(BB)_o \qquad (1)$$

The total number of monomers present is

$$N_o = (AA)_o + (BB)_o = (AA)_o(1 + \alpha/\beta) \qquad (2)$$

Define

$$\Phi = \alpha + \beta - 2\alpha\beta \qquad (3)$$

then the frequency function is

$$F(r) = (1 - \alpha)^2 \alpha^{r-1} \beta^r / \Phi$$

for $\underline{r}$ $\underline{AA}$ units, r - 1 $\underline{BB}$ units

$$+ 2(1 - \alpha)(1 - \beta)\alpha^r \beta^r / \Phi$$

for $\underline{r}$ $\underline{AA}$ units, $\underline{r}$ $\underline{BB}$ units

$$+ (1 - \beta)^2 \alpha^{r+1} \beta^r / \Phi$$

for $\underline{r}$ $\underline{AA}$ units, r + 1 $\underline{BB}$ units

$$+ (1 - \beta)^2 \alpha / \Phi \qquad 1 \leqslant r < \infty \tag{4}$$

for unreacted $\underline{BB}$ units

(Case, 2). Define

$$\Lambda = \alpha W_{BB} + \beta W_{AA} \tag{5}$$

The weight distribution is

$$W(r) = (1 - \alpha)^2 \alpha^{r-1} \beta^r \{rW_{AA} + (r - 1)W_{BB}\} / \Lambda$$

for $\underline{r}$ $\underline{AA}$ units, r - 1 $\underline{BB}$ units

$$+ 2(1 - \alpha)(1 - \beta)\alpha^r \beta^r \{rW_{AA} + rW_{BB}\} / \Lambda$$

for $\underline{r}$ $\underline{AA}$ units, $\underline{r}$ $\underline{BB}$ units

$$+ (1 - \beta)^2 \alpha^{r+1} \beta^r \{rW_{AA} + (r + 1)W_{BB}\} / \Lambda$$

for $\underline{r}$ $\underline{AA}$ units, r + 1 $\underline{BB}$ units

$$+ (1 - \beta)^2 \alpha W_{BB} / \Lambda \qquad 1 \leqslant r \leqslant \infty \tag{6}$$

for unreacted $\underline{BB}$ units

(Case, 2). Alternate equations for F(r) and W(r) are given by Case (2). Grethlein (3) presents more complex equations in which compensation for by-product elimination is made.

The average molecular weights are

$$\overline{M}_n = \frac{\beta W_{AA} + \alpha W_{BB}}{\alpha + \beta - 2\alpha\beta} = \frac{\Lambda}{\Phi} \tag{7}$$

$$\overline{M}_w = \frac{1 + \alpha\beta}{1 - \alpha\beta}\left[\frac{\beta W_{AA}^2 + \alpha W_{BB}^2}{\beta W_{AA} + \alpha W_{BB}}\right] + \frac{4\alpha\beta W_{AA} W_{BB}}{(1 - \alpha\beta)(\beta W_{AA} + \alpha W_{BB})} \tag{8}$$

Now if $\alpha = \beta$, $W_{AA} = W_{BB} = W$, these formulas reduce to the Schulz-Flory distribution,

$$\overline{r}_n = \overline{M}_n/W = 1/(1 - \alpha) \tag{9}$$

$$\overline{r}_w = \overline{M}_w/W = (1 + \alpha)/(1 - \alpha) \tag{10}$$

It is instructive to examine the distribution where one monomer is in excess and the other has undergone 100% reaction, as is done in the synthesis of oligomers (4).

Let $\alpha = 1$; then $\beta = (AA)_o/(BB)_o < 1$

$$F(r) = (1 - \beta)\beta^r \qquad 0 \leqslant r \leqslant \infty \tag{11}$$

where $\underline{r}$ is the number of $\underline{AA}$ units per molecule. The weight distribution is

$$W(r) = W_r F(r) / \sum_{r=0}^{\infty} W_r F(r) \tag{12}$$

where W_r is the exact molecular weight of each species $\underline{r}$.

Equation (11) shows that unless β is very small, less than 0.1, there will be significant portions of the oligomers with $r > 1$.

The distribution functions (4) and (6) are very similar to the Schulz-Flory function, except for the constants in each term. Therefore curves of these functions are not presented.

3. The Principle of Equal Reactivity, and Deviations from It

Consider the polyester reaction in the presence of a heavy metal catalyst

$$2H[OCH_2CH_2O\overset{O}{\overset{\|}{C}}C_6H_4\overset{O}{\overset{\|}{C}}]_1OCH_2CH_2OH$$

$$\rightarrow H[OCH_2CH_2O\overset{O}{\overset{\|}{C}}C_6H_4\overset{O}{\overset{\|}{C}}]_2OCH_2CH_2OH + HOCH_2CH_2OH \qquad (1)$$

in which an ethylene glycol unit is split out as by-product This equation can be generalized to the reesterification equation

$$R_r + R_s \rightleftharpoons R_{r+s-t} + R_t \qquad t \geqslant 0 \qquad (2)$$

where the index is the number of terephthalate units in the chain. This equation results when we assume the principle of equal reactivity for all molecules and all ester groups. Equation (1) can be written in the form

$$\overset{O}{\overset{\|}{\text{RCOR}'}} + \text{HOR}'' \rightleftharpoons \overset{O}{\overset{\|}{\text{RCOR}''}} + \text{HOR}' \tag{3}$$

If this reaction is independent of R, R', and R'' (for the present, we restrict R, R', and R'' to units other than hydrogen), then the transesterification equilibrium constant must be unity, because the reaction is an ester interchange reaction in both the forward and the reverse directions.

If we now add the esterification reaction

$$\overset{O}{\overset{\|}{\text{RCOH}}} + \text{HOR}'' \rightleftharpoons \overset{O}{\overset{\|}{\text{RCOR}''}} + \text{HOH} \tag{4}$$

to the transesterification scheme and define R_r as molecules with two glycol ends, Q_r as molecules with one acid end and one glycol end, and P_r as molecules with two acid function ends and further assume the principle of equal reactivity, then

$$R_r = R_1\alpha^{r-1} \qquad Q_r = Q_1\alpha^{r-1} \qquad P_r = P_1\alpha^{r-1} \tag{5}$$

Moreover, (1) $\alpha = R_1/G = P_1/G = Q_1/2G$ (6)

(2) equilibrium constant for esterification is unity

(3) concentration of water equals concentration of glycol

Hence

$$P_r = R_r = Q_r/2 = G\alpha^r \tag{7}$$

where $\underline{G}$ is the residual glycol or water concentration. This means that by assuming equal reactivity we must remove the parenthetical restriction on equation (3) because the esterification and transesterification reactions are equivalent.

Challa (5) has demonstrated that the equilibrium constant of transesterification increases as the molecular weight of the equilibrium polymer increases. He interprets this to mean that the equilibrium constant for monomer is different from that of the rest of the polymer. With this interpretation, he writes

$$R_1 + R_1 \rightleftharpoons R_2 + G \qquad K_{11} \tag{8}$$

$$R_1 + R_r \rightleftharpoons R_{r+1} + G \qquad K_1 \qquad r \neq 1 \tag{9}$$

$$R_r + R_s \rightleftharpoons R_{r+s} + G \qquad K_o \qquad r \neq 1,\ s \neq 1 \tag{10}$$

The molecular weight distribution is easily shown to be be

$$R_r = \frac{K_{11}R_1^{\ 2}}{G}\left[\frac{K_1 R_1}{G}\right]^{r-2} \qquad r > 1 \tag{11}$$

(Challa, 5). Substitution of (11) into the equilibrium equation obtained from (10) shows that

$$K_1^{\ 2} = K_o K_{11} \tag{12}$$

With the exception of this case, the preceding examples are based on the principle of equal reactivity. Certainly, as *r* becomes large, the reactivity of the larger molecules must approach each other because as the chain gets longer and longer, the influence of chain length on the reactivity of the terminal hydroxyl group or the terminal ester group is negligible. Thus deviation from the principle must occur only for low molecular weight molecules: the oligomers. On this basis, one should expect the equilibrium constant to vary as shown in Figure 4.3.1; the equilibrium constant may either increase or decrease as *r* is increased, but after some arbitrary value of *r*, say *a*, all of the remaining values of K_T are essentially independent of *r*. For this situation, the molecular weight distribution must be of the form

$$R_r = K'(K_T R_1/G)^{r-a} \qquad r \geqslant a \tag{13}$$

where *a* is the smallest value of *r* that does not deviate in reactivity from the larger members, and

$$R_r = f(R_1, G, K_r, K_1) \qquad r < a \tag{14}$$

It is important to note here that K_T is not equal to unity, and the distribution has the same form as the Schulz-Flory equation, except that the definition of the parameters is altered. If the weight fraction of molecules which deviate from the principle is small compared to the total weight of the system, then the bulk of the molecules still follow

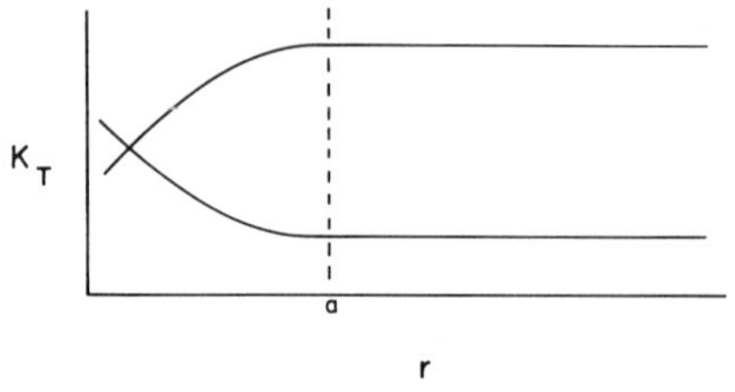

Figure 4.3.1 - Hypothetical variation of the transesterification equilibrium constant with chain length; above r = a, the constant does not vary with *r*.

a geometric type of distribution. We can take as an idealized case the Schulz-Flory equation (1.2.2) to calculate the value of r = a at a weight fraction of 0.01.

$$\int_0^a W(r)\,dr = (1/p) - [1 + (1 - p)a]p^{a-1} = 0.01 \qquad (15)$$

If we let $\bar{r}_n = 1/(1-p) = 80$, a reasonable value for polyesters in the useful textile range, then a = 12. Now, it is inconceivable that the reactivity of an end group can be influenced by groups in the molecule which are 12 repeat units removed unless there exists some type of electrical interaction not normally considered to affect the reactivity of the usual organic monomers.

On the other hand, let us for the moment suppose that the specific rate constant for condensation of an *AB* type polymer does depend on the size of the reacting molecules in the following fashion:

$$P_s + P_t = P_r$$

$$k_r = k(1 - r\xi); \quad r = s + t \tag{16}$$

where ξ can be negative, zero, or positive depending on whether k_r increases, is independent, or decreases as the molecular size increases. This equation can be substituted into the differential equation for molecular size as a function of time. The resultant frequency function is

$$F(r) = B^{(1-2\xi r)}(1 - B)^{r-1}G(r) \tag{17}$$

(Nanda and Jain, 6) where

$$B = 2\xi/[2\xi - 1 + \exp\{\xi N_o kt\}] \tag{18}$$

N_o = initial concentration of monomer

$$G(r) = \left[\frac{1 - \xi r}{r - 1}\right] \sum_{s=1}^{r-1} G(r - s)G(s) \qquad r \geqslant 2 \tag{19}$$

$$G(1) = 1 \tag{20}$$

$$\bar{r}_n = [1 - (1 - 2\xi) \exp(-\xi N_o kt)]/2\xi \tag{21}$$

$$\bar{r}_w = [1 - (1 - 2\xi) \exp(-2\xi N_o kt)]/2\xi \tag{22}$$

Figures 4.3.2 and 4.3.3 show the frequency and weight distributions as a function of $\underline{r}$ for $\bar{r}_n = 100$, $\xi = \pm 0.002$.

Curve	ξ	$\bar{r}_w$
a	0	199 - the Schulz-Flory distribution
b	+0.002	160
c	-0.002	238

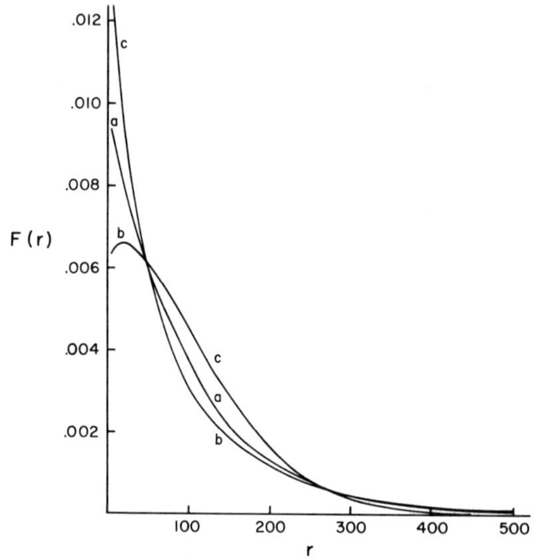

Figure 4.3.2 - The frequency distribution as a function of r for an AB type condensation polymer when k_p varies linearly with r, Equation (16), and $\bar{r}_n = 100$ (after Nanda and Jain, 6).

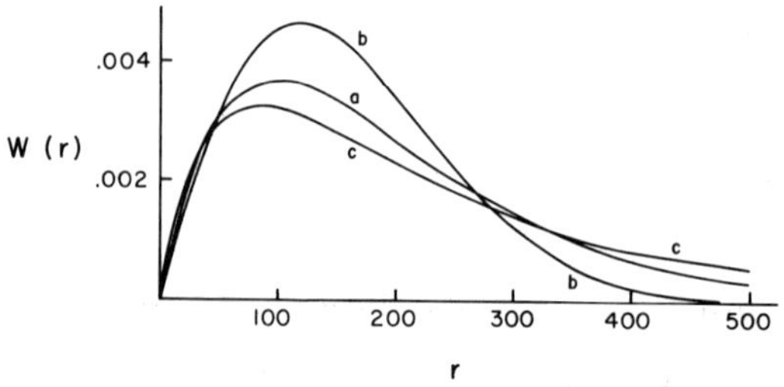

Figure 4.3.3 - The weight distribution as a function of r when k_p varies linearly with r. Same conditions as in Figure 4.3.2.

4. Other Simple Linear Condensation Cases

A. AA reacting with BC. BC is a cyclic monomer (anhydride); within a given molecule of BC, group B must react before group C.

Let α, β, γ be, respectively, the fractions of A, B, and C that have reacted, and let $\beta > \gamma$. The conservation conditions are

$$(BC)_o(\beta + \gamma) = 2\alpha(AA)_o \quad (1)$$

$$N_o = (AA)_o + (BC)_o = (AA)_o[1 + 2\alpha/(\beta + \gamma)] \quad (2)$$

Define

$$\eta = 2\gamma/(\beta + \gamma) \quad (3)$$

$$\Phi = 1 - 2\alpha + [2\alpha/(\beta + \gamma)] \quad (4)$$

The frequency function is

$$F(r) = (1 - \alpha)^2(\alpha\eta)^{r-1}/\Phi$$

for r AA units, r - 1 BC units

$$+ 2\alpha(1 - \alpha)(1 - \eta)(\alpha\eta)^{r-1}/\Phi$$

for r AA units, r BC units

$$+ (1 - \eta)^2\alpha^2(\alpha\eta)^{r-1}/\Phi$$

for r AA units, r + 1 BC units

$$+ 2\alpha(1 - \beta)/(\beta + \gamma)\Phi \qquad r \geq 1 \quad (5)$$

for unreacted BC units

(Case, 2). Define

$$\Lambda = W_{AA} + [2\alpha/(\beta + \gamma)]W_{BC} \tag{6}$$

Then

$$W(r) = (1 - \alpha)^2(\alpha\eta)^{r-1}\{rW_{AA} + (r - 1)W_{BC}\}/\Lambda$$

$$+ 2\alpha(1 - \alpha)(1 - \eta)(\alpha\eta)^{r-1}\{rW_{AA} + rW_{BC}\}/\Lambda$$

$$+ (1 - \eta)^2\alpha^2 (\alpha\eta)^{r-1}\{rW_{AA} + (r + 1)W_{BC}\}/\Lambda$$

$$+ 2\alpha(1 - \beta)W_{BC}/(\beta + \gamma)\Lambda \qquad r \geq 1 \tag{7}$$

(Case, 2) where the terms in W(r) are in the same order as given in F(r).

The average molecular weights are

$$\overline{M}_n = \frac{(\beta + \gamma)W_{AA} + 2\alpha W_{BC}}{2\alpha + \beta + \gamma - 2\alpha(\beta + \gamma)} = \frac{\Lambda}{\Phi} \tag{8}$$

$$\overline{M}_w = \left\{W_{AA}^2\left[\frac{1 + \alpha\eta}{1 - \alpha\eta}\right] + W_{BC}^2\left[\frac{2\alpha}{\beta + \gamma} + \frac{2\alpha^2}{1 - \alpha\eta}\right]\right.$$

$$\left. + W_{AA}W_{BC}\left[\frac{4\alpha}{1 - \alpha\eta}\right]\right\}/\Lambda \tag{9}$$

B. AA reacting with BC. BC is an unsymmetrical acid or glycol.

Let α, β, and γ be the fractions of reacted A, B, and C, respectively. The conservation conditions are the same as in Section A, but without the restriction $\beta > \gamma$. Define

$$\eta = 2\beta\gamma/(\beta + \gamma) \tag{10}$$

$$\Phi = 1 - 2\alpha + 2\alpha/(\beta + \gamma) \tag{11}$$

Then

$$F(r) = (1 - \alpha)^2(\alpha\eta)^{r-1}/\Phi$$

for $\underline{r}$ $\underline{AA}$ units, $r - 1$ $\underline{BC}$ units

$$+ 2\alpha(1 - \alpha)(1 - \eta)(\alpha\eta)^{r-1}/\Phi$$

for $\underline{r}$ $\underline{AA}$ units, $\underline{r}$ $\underline{BC}$ units

$$+ (1 - \eta)^2\alpha^2(\alpha\eta)^{r-1}/\Phi$$

for $\underline{r}$ $\underline{AA}$ units, $r + 1$ $\underline{BC}$ units

$$+ 2\alpha(1 - \beta)(1 - \gamma)/(\beta + \gamma)\Phi \qquad r \geqslant 1 \tag{12}$$

for unreacted $\underline{BC}$ units

(Case, 2). This equation is formally the same as that in Section A except for the last term.

The equation in Section A for W(r) can be used here by using the definitions of Λ given there and η given here, except that the last term must be replaced by

$$2\alpha(1 - \beta)(1 - \gamma)W_{BC}/(\beta + \gamma)\Lambda$$

$\overline{M}_n$ is the same as given in Equation (8).

$$\overline{M}_w = \{W_{AA}{}^2(1 + \alpha\eta) + W_{BC}{}^2[2\alpha(\beta + \gamma + \alpha\beta^2 + \alpha\gamma^2)$$

$$\div (\beta + \gamma)^2] + 4\alpha W_{AA}W_{BC}\}/(1 - \alpha\eta)\Lambda \tag{13}$$

The effect of a different reactivity of A with B and C can be demonstrated by letting the rate constant for the reaction of A with B be k and that for the reaction of A with C be k', and by letting $\varkappa = k'/k$. Furthermore, let A_o molecules of AA react with B_o molecules of BC and require that $B_o > A_o$ so that all molecules are capped with BC molecules. At the end of the reaction a fraction x of BC does not react at all. This quantity can be found from the equation

$$(\varkappa - 1)x + \varkappa x^{\frac{1}{2}} = (2\varkappa - 1)[1 - (A_o/B_o)] \tag{14}$$

(Hermans, 7). The degree of polymerization of molecules containing AA units is

$$\bar{r}_n = A_o/(B_o - A_o - x) \tag{15}$$

C. AB reacting with C and itself. C is a capping agent. B and C react only with A.

Let ν = the ratio of C groups to B groups

$$= (\alpha - \beta)/\gamma \tag{16}$$

$$N_o/(1 + \nu) = \text{the number of AB units} \tag{17}$$

The number of unreacted C units $= N_o\nu(1 - \nu)/(1 + \nu)$ (18)

$$F(r) = (1 - \alpha)(\nu\gamma + 1 - \alpha)\beta^{r-1}/(1 + \nu - \alpha)$$

for r AB units, 0 or 1 C units

$$+ \nu(1 - \gamma)/(1 + \nu - \alpha) \quad (19)$$

for unreacted C units

Let $\Lambda = W_{AA} + W_C$ (20)

$$W(r) = (1 - \beta)(\nu\gamma)\beta^{r-1}(rW_{AB} + W_C)/\Lambda$$

for r AB units, 1 C unit

$$+ (1 - \beta)(1 - \alpha)\beta^{r-1}rW_{AB}/\Lambda$$

for r AB units, 0 C unit

$$+ \nu(1 - \gamma)W_C/\Lambda \quad (21)$$

for unreacted C units

(Case, 2).

$$\overline{M}_n = (W_{AB} + \nu W_C)/(1 + \nu - \alpha) \quad (22)$$

$$\overline{M}_w = [W_{AB}^2(1 + \beta)/(1 - \beta) + 2W_{AB}W_C(\alpha - \beta)/(1 - \beta)$$

$$+ \nu W_C^{\,2}]/\Lambda \quad (23)$$

D. AA reacting with BB and C. C is a capping agent which can only react with A.

Let ν be the ratio of C to B units

$$(BB)_o = 2\alpha(AA)_o/(2\beta + \nu\gamma) \quad (24)$$

$$(C)_o - \nu(BB)_o \quad (25)$$

The number of unreacted C units is

$$N_o 2\alpha\nu(1 - \gamma)/(2\beta + \gamma)\{1 + [2\alpha(1 + \nu)/(2\beta + \nu\gamma)]\} \quad (26)$$

Let $\Phi = 1 - 2\alpha + [2\alpha(1 + \nu)/(2\beta + \nu\gamma)]$ (27)

$$\eta = \alpha/(2\beta + \nu\gamma) \quad (28)$$

$$\zeta = 2\eta\beta^2 \quad (29)$$

Then

$F(r) = 2\eta(1 - \beta)^2 \zeta^{r-1}/\Phi$

for r - 1 AA units, r BB units

$+ (1 - \alpha)^2 \zeta^{r-1}/\Phi$

for r AA units, r - 1 BB units

$+ \eta^2 \nu^2 \gamma^2 \zeta^{r-1}/\Phi$

for r AA units, r - 1 BB units, 2 C units

$+ 4\eta\beta(1 - \alpha)(1 - \beta)\zeta^{r-1}/\Phi$

for r AA units, r BB units

$+ 4\eta^2\beta\nu\gamma(1 - \beta)\zeta^{r-1}/\Phi$

for r AA units, r BB units, 1 C unit

$+ 2\eta\nu\gamma(1 - \alpha)\zeta^{r-1}/\Phi$

for r AA units, r - 1 BB units, 1 C unit

$+ 2\eta\nu(1 - \gamma)/\Phi \quad r \geqslant 1$ (30)

for unreacted C units

(Case, 2). Now define

$$\Lambda = W_{AA} + 2\eta W_{BB} + 2\eta\nu W_C \quad (31)$$

Then

$$W(r) = 2\eta(1 - \beta)^2\zeta^{r-1}\{(r - 1)W_{AA} + rW_{BB}\}/\Lambda$$

$$+ (1 - \alpha)^2\zeta^{r-1}\{rW_{AA} + (r - 1)W_{BB}\}/\Lambda$$

$$+ \eta^2\nu^2\gamma^2\zeta^{r-1}\{rW_{AA} + (r - 1)W_{BB} + 2W_C\}/\Lambda$$

$$+ 4\eta\beta(1 - \alpha)(1 - \beta)\zeta^{r-1}\{rW_{AA} + rW_{BB}\}/\Lambda$$

$$+ 4\eta^2\beta\nu\gamma(1 - \beta)\zeta^{r-1}\{rW_{AA} + rW_{BB} + W_C\}/\Lambda$$

$$+ 2\eta\nu\gamma(1 - \alpha)\zeta^{r-1}\{rW_{AA} + (r - 1)W_{BB} + W_C\}/\Lambda$$

$$+ 2\eta\nu(1 - \gamma)W_C/\Lambda \qquad (32)$$

(Case, 2).

$$\overline{M}_n = \Lambda/\Phi \qquad (33)$$

E. AA reacting with BC. A and B react with C.

Let ν be the ratio of BC to AA units. The conservation conditions are

$$2\alpha + \nu\beta = \nu\gamma, \qquad N_o = (1 + \nu)(AA)_o \qquad (34)$$

Define

$$\Phi = 1 + \nu(1 - \gamma) \qquad (35)$$

$$F(r) = (1 - \alpha)^2/\Phi$$

for unreacted AA units

$$+ 2\alpha(1 - \alpha)(1 - \beta)\beta^{r-1}/\Phi$$

for r BC units (r > 0), 1 AA unit at the end

$$+ \nu(1 - \gamma)(1 - \beta)\beta^{r-1}/\Phi$$

for r BC units (r > 0)

$$+ \alpha^2(1 - \beta)^2\beta^{r-2}(r - 1)/\Phi \quad (36)$$

for r BC units (r > 1), 1 AA unit in chain

(Case, 2). Define

$$\Lambda = W_{AA} + \nu W_{BC} \quad (37)$$

$$W(r) = (1 - \alpha)^2 W_{AA}/\Lambda$$

$$+ 2\alpha(1 - \alpha)(1 - \beta)\beta^{r-1}(rW_{BC} + W_{AA})/\Lambda$$

$$+ \nu(1 - \gamma)(1 - \beta)\beta^{r-1} rW_{BC}/\Lambda$$

$$+ \alpha^2(1 - \beta)^2\beta^{r-2}(r - 1)(rW_{BC} + W_{AA})/\Lambda \quad (38)$$

(Case, 2).

$$\overline{M}_n = \Lambda/\Phi \quad (39)$$

$$\overline{M}_w = \{W_{AA}{}^2 + 4\alpha W_{AA}W_{BC}/(1 - \beta)$$

$$+ W_{BC}{}^2[\nu + 2\alpha(\alpha + 2\beta)/(1 - \beta)^2]\}/\Lambda \quad (40)$$

F. AB reacts with CC or CD. Kinetics.

The rate of reaction of A depends upon whether or not

B has reacted. Likewise, rate of reaction of C depends upon whether or not B or D has reacted. To evaluate the distribution, the rate constants must be known. An example is given in reference (8).

5. Further Polymerization of Polymers with an Initial Geometric Distribution

A. Further polymerization of AB when the initial distribution is geometric.

This is a consideration of what would happen if a polymerization were interrupted and then continued again. Let the initial distribution be

$$F^{o}(r) = (1 - \alpha)\alpha^{r-1} \tag{1}$$

where α is the extent of initial reaction. If α' is the extent of reaction for the second step, then

$$F(r) = (1 - \alpha)(1 - \alpha')[\alpha + \alpha'(1 - \alpha)]^{r-1} \tag{2}$$

$$= (1 - \epsilon)\epsilon^{r-1} \tag{3}$$

(Hermans, 9) where

$$\epsilon = \alpha + \alpha'(1 - \alpha) \tag{4}$$

so the Schulz-Flory distribution is maintained. Reequilibrium reactions such as

$$(AB)_i + (AB)_j \rightleftharpoons (AB)_k + (AB)_{i+j-k} \qquad k > 0 \tag{5}$$

are assumed not to occur in the derivation.

B. Further polymerization of AB when the initial distribution is a superposition of two geometric distributions

Again, as in Section A, reequilibrium reactions are assumed not to occur.

Let

$$P_r^o = a\alpha^{r-1} + b\beta^{r-1} \tag{6}$$

and

$$s_o = \sum_{r=1}^{\infty} P_r^o = \frac{a}{1 - \alpha} + \frac{b}{1 - \beta} \tag{7}$$

Then

$$F^o(r) = P_r^o/s_o \tag{8}$$

If α' is the extent of conversion for the second step,

$$\alpha' = (s_o - s)/s_o \tag{9}$$

Then

$$P_r = A\lambda^{r-1} + B\mu^{r-1} \tag{10}$$

(Hermans, 9) where λ and μ are the roots of the equation

$$s_o^2 x^2 - [(\alpha + \beta)s_o^2 + (s_o - s)(a + b)]x$$

$$+ [\alpha\beta s_o^2 + (s_o - s)(a\beta + b\alpha)] = 0 \tag{11}$$

$\underline{A}$ and $\underline{B}$ can be derived from the equations

$$A + B = s^2(a + b)/s_o^2 \tag{12}$$

$$\lambda A + \mu B = (s^2/s_o^2)(a\alpha + b\beta)$$

$$+ s^2(s_o - s)(a + b)^2/s_o^4 \tag{13}$$

C. Further polymerization of $\underline{AA}$ with $\underline{BB}$ when the initial distribution of both is geometric.

The initial distribution of $\underline{AA}$ is $P_r^{oa} = a\alpha^{r-1}$ and of $\underline{BB}$ is $P_r^{ob} = b\beta^{r-1}$, type $\underline{BB}$ molecules are in excess, that is, $\Sigma P_r^{oa} < \Sigma P_r^{ob}$, and the reaction has gone to completion, all molecules having $\underline{B}$'s at the end. Then

$$P_r = \left[\frac{b(1 - \alpha) - a(1 - \beta)}{b(1 - \alpha)}\right]^2 b[A\rho^{r-1} + B\sigma^{r-1}] \tag{14}$$

(Hermans, 9) where ρ and σ are the roots of

$$\xi^2 - [\alpha + \beta + a(1 - \beta)^2/b]\xi + \alpha\beta = 0 \tag{15}$$

and $\underline{A}$ and $\underline{B}$ must be found from

$$A + B = 1; \qquad \rho A + \sigma B = \beta + a(1 - \beta)^2/b \tag{16}$$

This treatment can be extended to the reaction of $\underline{k}$ geometric distributions of type $\underline{AA}$ molecules with $\underline{j}$

geometric distributions of type BB molecules (Hermans, 9).

6. Copolymerization of Condensation Polymers AB and CD. A reacts with B and C. B reacts with A and D.

When two or more monomers are mixed to form copolymers, the various groups may differ in their reactivity, provided that size of the molecules cannot influence the reactivity. In the strictest interpretation of the theory this means that ethylene glycol has the same reactivity as 1, 10 decanediol. One may think that this interpretation can be avoided by assuming that the unreacted monomers can be measured and the total concentration of a particular end group (say carboxyl groups) or the molecular weight of the resultant polymer is obtained so that the individual extents of reaction can be calculated. This procedure uses the principle of equal reactivity to calculate parameters which are then used in equations derived on the basis of this principle; it cannot be used to show differences in monomer reactivity. The influence of monomer reactivity on the distribution was discussed in Section 3. Similar procedures could be applied to copolymers.

We present here only the equations for two hydroxy acids or similar materials AB and CD.

Let ν be the ratio of CD to AB units. The conservation condition is $\alpha + \nu\gamma = \beta + \nu\delta$. Define the following term:

$$\psi(r, i) = (\alpha\beta/\nu\gamma\delta)^{i}[\nu\gamma\delta/(\alpha + \nu\gamma)]^{r} r!/i!(r - i)!$$

$$\div [1 + \nu(\alpha + \nu\gamma)](\alpha + \nu\gamma) \tag{1}$$

where $\underline{r}$ is the number of units in the chain, not including the two end units, and $\underline{i}$ is the number of AB units in this chain.

$$
\begin{aligned}
F(r, i) &= \beta(1 - \alpha)\psi(r, i)\alpha(1 - \beta) && \text{for } r + 2 \text{ units, A and B groups at the ends, } i + 2 \text{ AB units} \\
&+ \beta(1 - \alpha)\psi(r, i)\nu\gamma(1 - \delta) && \text{for } r + 2 \text{ units, A and D groups at the ends, } i + 1 \text{ AB units} \\
&+ \nu\delta(1 - \gamma)\psi(r, i)\alpha(1 - \beta) && \text{for } r + 2 \text{ units, B and C groups at the ends, } i + 1 \text{ AB units} \\
&+ \nu\delta(1 - \gamma)\psi(r, i)\nu\gamma(1 - \delta) && \text{for } r + 2 \text{ units, C and D groups at the ends, } i \text{ AB units} \\
&+ (1 - \alpha)(1 - \beta)/[1 + \nu - (\alpha + \nu\gamma)] && \text{for unreacted AB units} \\
&+ \nu(1 - \gamma)(1 - \delta)/[1 + \nu - (\alpha + \nu\gamma)] \quad 0 \leqslant r \leqslant \infty && \text{for unreacted CD units}
\end{aligned} \tag{2}
$$

(Case, 2). Define

$$\Omega(r, i) = (\alpha\beta/\nu\gamma\delta)^{i}[\nu\gamma\delta\ /(\alpha + \nu\gamma)]^{r} r!/i!(r - i)!$$
$$\div (W_{AB} + \nu W_{CD})(\alpha + \nu\gamma) \tag{3}$$

$$
\begin{aligned}
&W(r, i) \\
&= \Omega(r, i)\{\beta(1 - \alpha)\alpha(1 - \beta)[(i + 2)W_{AB} + (r - i)W_{CD}] \\
&+ \beta(1 - \alpha)\nu\gamma(1 - \delta)[(i + 1)W_{AB} + (r - i + 1)W_{CD}] \\
&+ \nu\delta(1 - \gamma)\alpha(1 - \beta)[(i + 1)W_{AB} + (r - i + 1)W_{CD}]
\end{aligned}
$$

$$+ \nu\delta(1 - \gamma)\nu\gamma(1 - \delta)[iW_{AB} + (r - i + 2)W_{CD}]\}$$

$$+ \frac{(1 - \alpha)(1 - \beta)W_{AB} + \nu(1 - \gamma)(1 - \delta)W_{CD}}{W_{AB} + \nu W_{CD}} \tag{4}$$

(Case, 2) where the terms are given in the same order as in F(r, i)

$$\overline{M}_n = (W_{AB} + \nu W_{CD})/[1 + \nu - (\alpha + \nu\gamma)] \tag{5}$$

Distributions for the following systems are given by Case (2):

AA reacting with BB and CC. B does not react with C.

AA and BB reacting with CC and DD. A and B react only with C and D and vice versa.

AA reacting with BC and DD. A reacts only with B, C, and D.

AA and DD reacting with BC. A and B react only with C and D and vice versa.

7. Coupled Polymers

An important means of producing polymers with novel properties is to react a prepolymer having one set of properties with a coupling agent which has different characteristics. An example is the spandex-type fiber which is composed of "soft segments" coupled with "hard segments." The soft segments are usually of two forms, either a Schulz-Flory type distribution of polyesters or a Poisson distribution of polyethers coupled with a hard segment of a high molecular weight diisocyanate. The distribution functions for coupled polymers are quite complex; we present here only the frequency distribution. If the frequency distribution is represented by F(n, j) where n is the number of monomers and j is the number of coupling units in a molecule of size n, j, then the weight fraction distribution is

$$W(n, j) = (nW_M + jW_C)F(n, j)/\overline{M}_n$$

where W_M and W_C are the molecular weights of the monomer repeat unit and coupling repeat unit, respectively. Examples have been calculated for only two cases, where the prepolymer is of the Schulz-Flory or the Poisson type. For each type of prepolymer we consider two additional examples: one in which we essentially double the molecular weight of the prepolymer, the other in which the molecular weight of the prepolymer is raised by a factor of about five. As one might expect, the distribution resulting from a Schulz-Flory prepolymer is of the same form, whereas that resulting from a Poisson prepolymer approaches the Schulz-Flory distribution as the extent of coupling is increased.

A. AB polymerized to extent of reaction α, then coupled with CC.

Let α = fraction of A that has reacted with B prior to coupling

α' = fraction of a previously unreacted A reacting with C

β' = fraction of a previously unreacted B reacting with C

γ = fraction of C that reacts

r = total number of AB units in the coupled polymer

j = total number of CC units

Define

$$\psi(r, j) = \frac{\left[\frac{2\gamma\alpha'\beta'}{(\alpha' + \beta')}\right]^{j-1} (1 - \alpha)^j \alpha^{r-j} \binom{r-1}{j-1}}{[1 - 2\gamma + 2\gamma/(\alpha' + \beta')]} \tag{1}$$

Then

$$\begin{aligned}
F(r, j) &= (1 - \gamma)^2 \psi(r, j) \qquad \text{for } j \text{ CC units, } j - 1 \text{ AB chains} \\
&+ 2(1 - \gamma)\gamma[1 - 2\alpha'\beta'/(\alpha' + \beta')]\psi(r, j) \qquad \text{for } j \text{ CC units, } j \text{ AB chains} \\
&+ [1 - 2\alpha'\beta'/(\alpha' + \beta')]^2 \gamma^2 \psi(r, j) \qquad 1 \leqslant j \leqslant r, \\
&1 \leqslant r < \infty \qquad \text{for } j \text{ CC units, } j + 1 \text{ AB chains} \\
&+ \frac{2\gamma(1 - \alpha')(1 - \beta')(1 - \alpha)\alpha^{r-1}}{(\alpha' + \beta')[1 - 2\gamma + 2\gamma/(\alpha' + \beta')]} \qquad j = 0, \\
&1 \leqslant r < \infty \qquad \text{for unreacted AB chains}
\end{aligned} \tag{2}$$

(Case, 10).

Note that this equation is restricted to values of j equal to or less than r, r remaining finite. Thus it appears that molecules of the type

CC	$j = 1, r = 0$
CC-AB-CC	$j = 2, r = 1$
CC-AB-CC-AB-CC	$j = 3, r = 2$
etc.	etc.

are specifically excluded. However, if α and γ are large, close to unity, then the concentration of such molecules with $j \geqslant 2$ will be unimportant. The concentration of unreacted CC units can be found by setting $j = 1$ in the first term of $F(r, j)$ and summing over r to obtain

$$\text{concentration of unreacted CC units} = \frac{(1 - \gamma)^2}{1 - 2\gamma + \dfrac{2\gamma}{(\alpha' + \beta')}} \tag{3}$$

The distribution over r is found by summing over j

$$F(r) = \sum_{j=1}^{\infty} F(r, j)$$

$$= \frac{\left[1 - \dfrac{2\alpha'\beta'\gamma}{(\alpha' + \beta')}\right]^2 (1 - \alpha)\left[\alpha + \dfrac{(1 - \alpha)2\alpha'\beta'\gamma}{(\alpha' + \beta')}\right]^{r-1}}{[1 - 2\gamma + 2\gamma/(\alpha' + \beta')]}$$

$$1 \leqslant j \leqslant r, \qquad 1 \leqslant r \leqslant \infty$$

$$+ \frac{2\gamma(1 - \alpha')(1 - \beta')(1 - \alpha)\alpha^{r-1}}{(\alpha' + \beta')[1 - 2\gamma + 2\gamma/(\alpha' + \beta')]} \qquad j = 0, \quad 1 \leqslant r < \infty \tag{4}$$

(Case, 10). This is a distribution of the geometric type. The number-average molecular weight is

$$\bar{M}_n = \frac{W_{CC} + 2\gamma W_{AB}/(\alpha' + \beta')(1 - \alpha)}{1 - 2\gamma + 2\gamma/(\alpha' + \beta')} \tag{5}$$

Two examples of coupled polymers are presented with the following simplifying assumptions: (a) $W_{CC} = W_{AB}$, so that $W(n, j) = (n + j)F(n, j)/\bar{r}_n$, and (b) the number of CC molecules added exactly equals the number of prepolymer molecules, so that we can assume that $\gamma = \alpha' = \beta'$. By letting the number-average degree of polymerization of the prepolymer be either 19 or 49, and the coupling reaction proceed until $\bar{r}_n = 100$, the values of α and γ can be determined. The results are presented in Figures 4.7.1 and 4.7.2 for $\bar{r}_n{}^o = 19$ and in 4.7.3 and 4.7.4 for $\bar{r}_n{}^o = 49$.

The frequency distribution for molecules containing *j* coupling agents is

$$F(j) = \sum_{r=1}^{\infty} F(r, j)$$

$$= \frac{[1 - 2\alpha'\beta'\gamma/(\alpha' + \beta')]^2 [2\alpha'\beta'\gamma/(\alpha' + \beta')]^{j-1}}{1 - 2\gamma + 2\gamma/(\alpha' + \beta')} \qquad j \geq 1$$

$$+ \frac{2\gamma(1 - \alpha')(1 - \beta')}{(\alpha' + \beta')[1 - 2\gamma + 2\gamma/(\alpha' + \beta')]} \qquad j = 0 \tag{6}$$

Thus

$$\bar{j}_n = 1/[1 - 2\gamma + 2\gamma/(\alpha' + \beta')] \tag{7}$$

for molecules that contain coupling agent.

The weight fraction of molecules of size *j* is given in Figure 4.7.5 for the two distributions $\bar{r}_n{}^o = 19$ and 49,

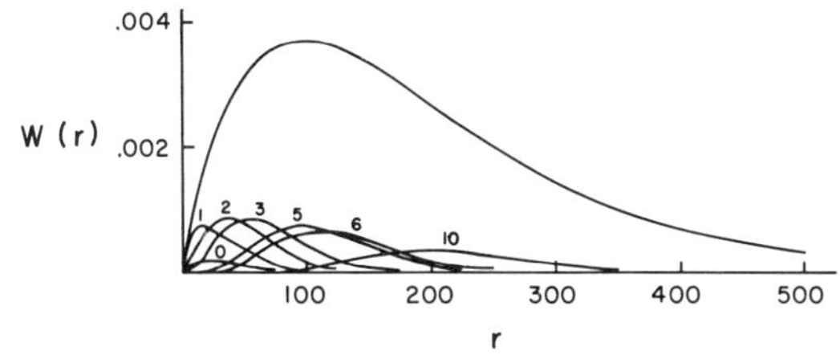

Figure 4.7.1 - Weight distribution of a Schulz-Flory type prepolymer with $\bar{r}_n^{\,o}$ = 19 coupled to make a polymer with $\bar{r}_n$ = 100, as a function of r, under the conditions $W_{AB} = W_{CC}$ and $\gamma = \alpha' = \beta'$. The overall distribution is the large curve, W(r); the smaller curves give the distribution of molecules W(r, j) containing j coupling units (after Case, 10).

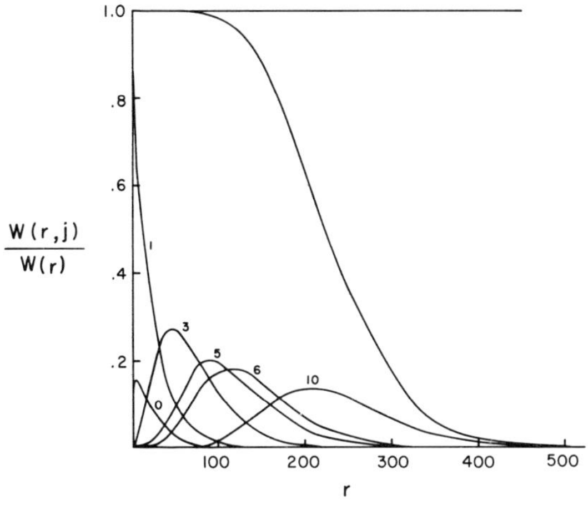

Figure 4.7.2 - Relative weight distribution of molecules containing j coupling units as a function of r for $\bar{r}_n^{\,o}$ = 19, $\bar{r}_n$ = 100, same assumptions are in Figure 4.7.1. The large curve gives $\sum_{j=0}^{10} W(j)$.

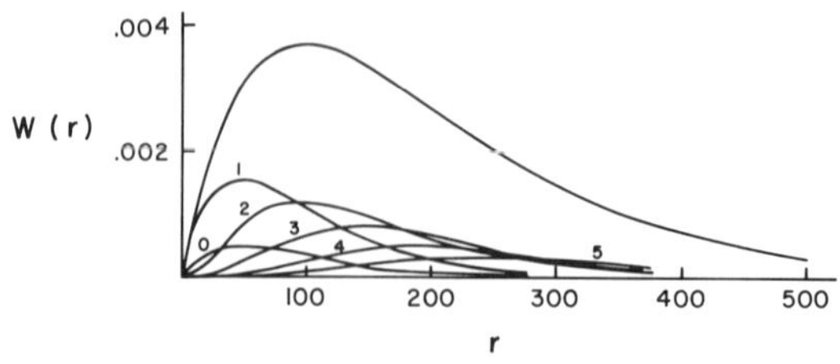

Figure 4.7.3 - Weight distribution ot a Schulz-Flory type prepolymer with $\bar{r}_n^{\,o}$ = 49 coupled to make a polymer with $\bar{r}_n$ = 100. Same assumptions as in Figure 4.7.1 (after Case, 10).

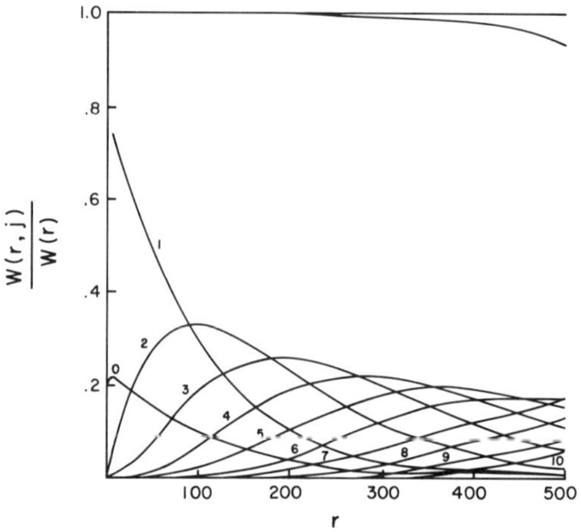

Figure 4.7.4 - Relative weight distribution of molecules containing j coupling units as a function of r for $\bar{r}_n^{\,o}$ = 4 $\bar{r}_n$ = 100. Same assumptions as in Figure 4.7.1. The curve in the upper right hand corner gives $\sum_{j=0}^{10} W(j)$.

$\bar{r}_n$ = 100. Note that $\bar{j}_n$ is not $\bar{r}_n / \bar{r}_n^{\,o}$; for the present distributions it is 4.95 and 1.87 respectively.

$$W(j) = \sum_{r=1}^{\infty} (r + j)F(r, j)/\bar{r}_n$$

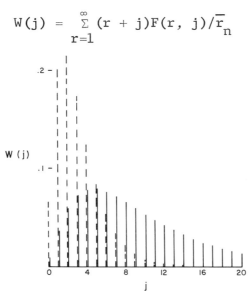

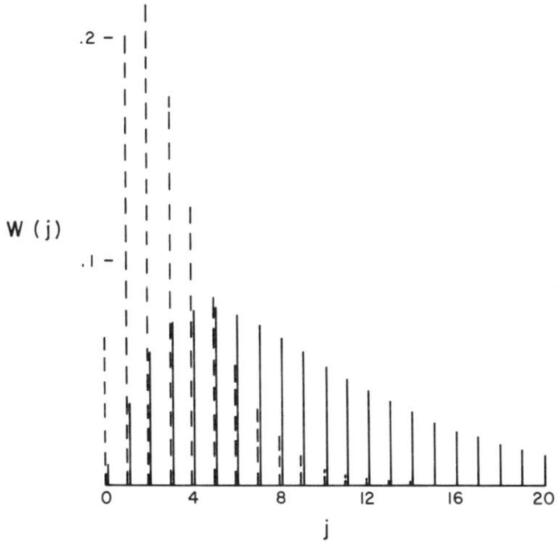

Figure 4.7.5 - Weight distribution of molecules W(j) containing j coupling units as a function of j. Broken line is for $\bar{r}_n^{\,o} = 49$, solid line for $\bar{r}_n^{\,o} = 19$, $\bar{r}_n = 100$ for both distributions, same assumptions as in Figure 4.7.1.

B. AB polymerized to extent of reaction α, then coupled with CD.

Let α = fraction of A that has reacted with B

α' = fraction of A, which has not reacted with B, but has reacted with C or D

γ = fraction of reacted C

δ = fraction of reacted D

r = total AB units in the polymer chain under consideration

j = number of poly AB prepolymer chains that originally existed before addition of CD to the chain under consideration

$$\psi(r, j) = \frac{\left[\frac{2\alpha'\gamma\delta}{\gamma + \delta}\right]^{j-1} (1 - \alpha)^j \alpha^{r-j} \binom{r-1}{j-1}}{1 - 2\alpha' + 2\alpha'/(\gamma + \delta)} \tag{8}$$

Then

$$F(r, j) = (1 - \alpha')^2 \psi(r, j) \qquad 2 \leqslant j \leqslant r, \quad 2 \leqslant r \leqslant \infty$$

$$+ 2(1 - \alpha')[1 - 2\gamma\delta/(\gamma + \delta)]\alpha' \psi(r, j)$$

$$+ [1 - 2\gamma\delta/(\gamma + \delta)]^2 \alpha'^2 \psi(r, j) \qquad \Bigg\} \; 1 \leqslant j \leqslant r, \quad 1 \leqslant r < \infty$$

$$+ (1 - \alpha')^2 (1 - \alpha)\alpha^{r-1}/[1 - 2\alpha' + 2\alpha'/(\gamma + \delta)] \qquad j = 1, \quad 1 \leqslant r < \infty$$

$$+ 2\alpha'(1 - \gamma)(1 - \delta)/(\gamma + \delta)[1 - 2\alpha' + 2\alpha'/(\gamma + \delta)] \tag{9}$$

unreacted CD units, $j = 0$

(Case, 10). The formula given by Case (10) sums to unity only if $\alpha' = \beta'$, where β' is defined as the fraction of B which has not reacted with A but has reacted with C or D.

$$\overline{M}_n = \left[\frac{W_{AB}}{1 - \alpha} + \left(\frac{2\alpha'}{\gamma + \delta}\right) W_{CD}\right] / [1 - 2\alpha' + 2\alpha'/(\gamma + \delta)] \tag{10}$$

C. AA and BB polymerized to extent of reaction α, then coupled with an excess of CC. A and B to react completely

Let α = fraction of A reacting with B

β = fraction of B reacting with A

γ = fraction of C reacting with A or B

m = the internal AA-BB pairs

i = number of prepolymer chains with both ends consisting of A groups in the molecule under consideration

j = number of prepolymer chains with one end A, one end B

k = number of prepolymer chains with both ends B

$\ell = i + j + k$

Furthermore, previously unreacted A and B groups must react completely with CC. The total number of i-type, j-type, and k-type molecules can be determined from Section 1.

$$F(M, i, j, k) = \frac{(1 - \gamma)\gamma^{\ell}\{\beta(1 - \alpha)^2\}^i\{\alpha(1 - \beta)^2\}^k}{(\alpha + \beta - 2\alpha\beta)^{\ell}}$$

$$\times \{2\alpha\beta(1 - \alpha)(1 - \beta)\}^j(\alpha\beta)^m\ell(m + \ell - 1)!/i!j!k!m!$$

for the conditions

$$0 \leqslant m < \infty \quad 0 < i < \ell \quad 1 \leqslant \ell < \infty$$

$$0 \leqslant j \leqslant \ell - i \quad k = \ell - i - j$$

$$+ (1 - \gamma) \quad \text{for } m = i = j = k = 0 \tag{11}$$

(Case, 10).

$$F(m, \ell) = \frac{(1 - \gamma)(\alpha\beta)^m\{\gamma(1 - \alpha\beta)\}^{\ell}(m + \ell - 1)!}{m!(\ell - 1)!}$$

$$0 \leqslant m < \infty \quad 1 \leqslant \ell < \infty$$

$$+ (1 - \gamma) \quad m = \ell = 0 \tag{12}$$

$$F(m + \ell) = (1 - \gamma)\gamma(1 - \alpha\beta)\{\alpha\beta + \gamma - \alpha\beta\gamma\}^{m+\ell-1}$$
$$0 \leqslant m + \ell - 1 < \infty$$
$$+ (1 - \gamma) \qquad m + \ell = 0 \tag{13}$$

(Case, 10)

$$\overline{M}_n = \{W_{CC} + \gamma[W_{AA} + (\alpha/\beta)W_{BB}]/[1 - 2\alpha + (\alpha/\beta)]\}/(1 - \gamma) \tag{14}$$

D. Poisson distribution of polymer of AA coupled with BC.

The Poisson prepolymer has a number-average degree of polymerization $\overline{r}_n = \nu + 1$ and a frequency distribution

$$F(r) = e^{-\nu}\nu^{r-1}/(r - 1)! \tag{15}$$

Let m = total AA units in the coupled polymer

j = total prepolymer chains

α = fraction of prepolymer chain ends that have reacted with BC

β = fraction of B reacted

γ = fraction of C reacted

$$\psi(m, j) = \frac{\{2\alpha\beta\gamma/(\beta + \gamma)\}^{j-1}e^{-j\nu}(j\nu)^{m-j}}{(m - j)![1 - 2\alpha + 2\alpha/(\beta + \gamma)]} \tag{16}$$

Then

$$F(m, j) = (1 - \alpha)^2\psi(m, j)$$

both ends AA units

$$+ 2(1 - \alpha)\alpha[1 - 2\beta\gamma/(\beta + \gamma)]\psi(m, j)$$

one end BC, other end AA

$$+ \alpha^2[1 - 2\beta\gamma/(\beta + \gamma)]\psi(m, j)$$

both ends BC

$$+ 2\alpha(1 - \beta)(1 - \gamma)/(\beta + \gamma)[1 - 2\alpha + 2\alpha/(\beta + \gamma)] \quad (17)$$

unreacted BC units

$$1 \leqslant j \leqslant m, \qquad 1 \leqslant m < \infty$$

(Case, 10).

$$\overline{M}_n = \frac{(\beta + \gamma)(\nu + 1)W_{AA} + 2\alpha W_{BC}}{2\alpha + \beta + \gamma - 2\alpha(\beta + \gamma)} \quad (18)$$

Figures 4.7.6 and 4.7.7 present the weight distribution for $\nu = 18$, $\alpha = \beta = \gamma = 0.9$, $\overline{r}_n = 100$, and Figure 4.7.8 presents that for $\nu = 48$, $\alpha = \beta = \gamma = 0.75$, $\overline{r}_n = 100$. As the amount of coupling increases, the distributions approach the Schulz-Flory distribution.

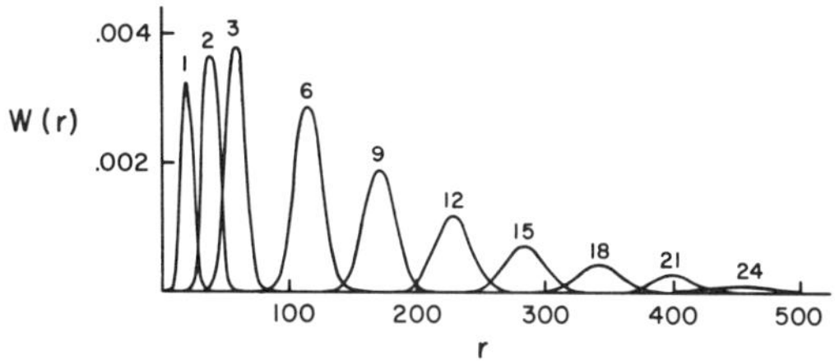

Figure 4.7.6 - Weight distribution of a Poisson type prepolymer with $\overline{r}_n^{\,o} = 19$ coupled to make a polymer with $\overline{r}_n = 100$ W(r, j) as a function of r under the conditions $W_{AA} = W_{BC}$ and $\alpha = \beta = \gamma$. Only every third subdistribution

is shown after j = 3. (After Case, 10).

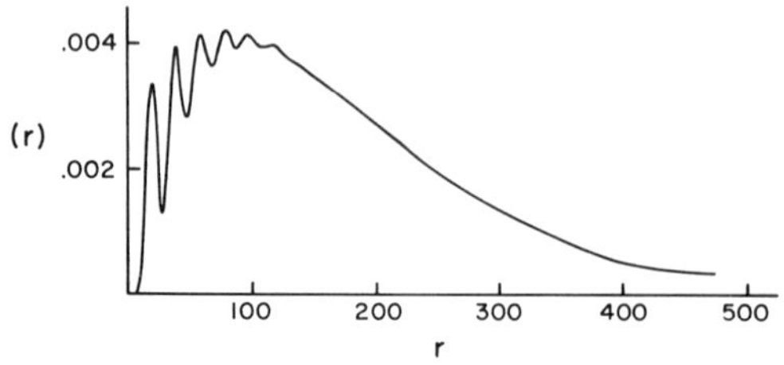

Figure 4.7.7 - Overall weight distribution for the curves given in Figure 4.7.6.

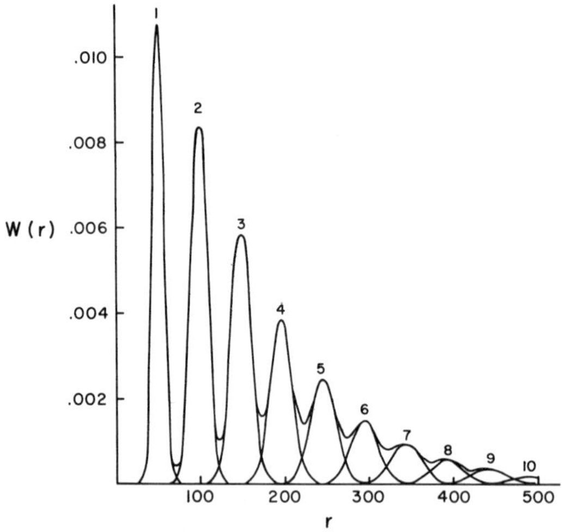

Figure 4.7.8 - Weight distribution of a Poisson type prepolymer with $\bar{r}_n^{\,o}$ = 49 coupled to make a polymer with $\bar{r}_n$ = 100, W(r, j) as a function of r. Same conditions as in Figure 4.7.6.

E. AB polymerized to extent of reaction α, then coupled with excess CC to extent of reaction γ, then recoupled with excess DD.

Let α = fraction of A that has reacted with B

γ = fraction of C that has reacted with A or B

δ = fraction of D that has reacted

m = number of AB units

j = number of original AB chains

k = number of coupled chains in the recoupled chain

ℓ = number of prepolymer CC units in the recoupled chain

Then

$$F(m, j, k, \ell)$$
$$= (1-\delta)\delta^{k+\ell}(1-\gamma)^{k+\ell}\gamma^{j}\alpha^{m-j}(1-\alpha)^{j}\binom{j-1}{k-1}\binom{k+\ell}{k}\binom{m-1}{j-1}$$
$$1 \leqslant k \leqslant j, \quad 1 \leqslant j \leqslant m, \quad 1 \leqslant m \leqslant \infty, \; 0 \leqslant \ell < \infty$$
$$+ (1-\delta)\delta^{\ell}(1-\gamma)^{\ell} \qquad 0 \leqslant \ell < \infty, \qquad j = k = m = 0 \tag{19}$$

(Case, 10).

$$\overline{M}_n = W_{DD} + \frac{\delta}{1-\gamma} W_{CC} + \frac{\gamma\delta}{(1-\alpha)(1-\gamma)} W_{AB} \tag{20}$$

F. Particularly narrow distributions via coupling reactions.

The following reactions are considered by Case (11).

I. AA and BB (great excess) → BBAABB, then remove excess BB

CC and BBAABB (great excess) → BBAABBCCBBAABB, then remove excess BBAABB, and continue in like manner

II. AA and 2BC → CBAABC
CBAABC and 2DE → EDCBAABCDE
etc.

III. AB and CD → ABCD
ABCD and EF → ABCDEF
etc.

G. Blocks of polymers of known distribution are coupled together.

The following reactions are considered by Case (12).

I. A series of Poisson type polymers coupled together.

II. Poisson type polymers coupled to Schulz-Flory type polymers.

III. A series of Schulz-Flory type polymers coupled together.

8. References

1. P. J. Flory, "Molecular Size Distribution in Linear Condensation Polymers," J. Am. Chem. Soc., 58, 1877 (1936).
2. L. C. Case, "Molecular Distributions in Polycondensations Involving Unlike Reactants. II. Linear Distributions," J. Polymer Sci., 29, 455 (1958).
3. H. E. Grethlein, "Exact Weight Fraction Distribution in Linear Condensation Polymerization," Ind. Eng. Chem. Fundamentals, 8, 206 (1969).

4. L. H. Peebles, Jr., M. W. Huffman, and C. T. Ablett, "Isolation and Identification of the Linear and Cyclic Oligomers of Poly(ethylene terephthalate) and the Mechanism of Cyclic Oligomer Formation," J. Polymer Sci., A1, 7, 479 (1969).
5. G. Challa, "The Formation of Polyethylene Terephthalate by Ester Interchange. I. The Polycondensation Equilibrium," Makromol. Chem., 38, 105 (1960).
6. V. S. Nanda and S. C. Jain, "Effect of Variation of the Bimolecular Rate Constant with Chain Length on the Statistical Character of Condensation Polymers," J. Chem. Phys., 49, 1318 (1968).
7. J. J. Hermans, private communication, 1963.
8. L. C. Case, "Molecular Distributions in Polycondensations Involving Unlike Reactants. VII. Treatment of Reactants Involving Nonindependent Groups," J. Polymer Sci., 48, 27 (1960).
9. J. J. Hermans, "Molecular Weight Distributions Resulting from Irreversible Polycondensation Reactions," Makromol. Chem., 87, 21 (1965).
10. L. C. Case, "Molecular Distributions in Polycondensations Involving Unlike Reactants. III. Distributions Arising in the Linear Coupling of Single Polymers," J. Polymer Sci., 37, 147 (1959).
11. L. C. Case, "Molecular Distributions in Polycondensations Involving Unlike Reactants. V. Particularly Narrow Distributions," J. Polymer Sci., 39, 175 (1959).

12. L. C. Case, "Molecular Distributions in Polycondensations Involving Unlike Reactants. VI. Independently Distributed Multivariate Distributions," *J. Polymer Sci.*, *39*, 183 (1959).

Chapter 5

Nonlinear Systems

Contents

1. Introduction

So far we have considered only the formation of linear molecules; the formation of branches and/or rings has been specifically excluded. Just as the distribution for copolymers is more complex than that for homopolymers, so also are the distributions which result when branching, ring formation, and gelation occur.

In vinyl-type polymerization, branch formation can occur by the transfer-to-polymer reaction

$$R_n^* + P_r \xrightarrow{k_{tr,p}} R_r^* + P_n \tag{1}$$

Now, the rate constant for this reaction is not simply $k_{tr,p}$, but depends also on the number of transfer sites available in the molecule P_r. In general, we assume that there is only one transfer site per repeat unit; hence the rate of reaction (1) is proportional to $rk_{tr,p}$.

Branch formation can also occur by polymerization through a terminal double bond:

$$R_n^* + P_r \xrightarrow{k_{p,p}} R_{r+n}^* \tag{2}$$

The terminal double bond may be formed either by the transfer-to-monomer reaction, which creates an unsaturated monomer radical that continues the chain

$$R_n^* + CH_2{=}CH\text{-}AH \xrightarrow{k_{tr,m}} P_nH + CH_2{=}CH\text{-}A^* \quad (3)$$

$$CH_2{=}CH\text{-}A^* + M \xrightarrow{k_p} CH_2{=}CH\text{-}A\sim\sim\sim\sim^* \quad (4)$$

(where A could be a group such as the acetate function) or by termination by disproportionation

$$R_n\text{-}CH_2^* + {}^*CH_2CH_2\text{-}R_s \xrightarrow{k_{t,d}} R_nCH_3 + R_sCH{=}CH_2 \quad (5)$$

These reactions can lead to a highly branched structure, but closed rings cannot be formed by this mechanism. It is generally assumed that gelation--the formation of insoluble, infusable, three-dimensional structures--is the result of extensive formation of cyclic structures. However, the reverse is not necessarily true: an "insoluble, infusable, three-dimensional structure" may be soluble in the proper solvent or may require an extended time to dissolve. A highly branched material may have limited solubility but not be cross-linked.

Cyclic as well as linear molecules can be formed during condensation polymerization whereby the "head" of a molecule condenses with its own "tail." Thus Nylon 6 and polyethylene terephthalate, inter alia, contain both linear and cyclic molecules.

At this point we introduce the concept of functionality. A vinyl group, CH_2=CH- or an AB-type condensation-type monomer (an ω-hydroxy acid) has a functionality of two; they can only form linear chains or simple rings.

A chain stopping agent or terminator, has a functionality of one. There are two types of trifunctional groups in condensation polymerization. Consider the system AB + RA_n where A can only react with B and vice versa and n is 3 or greater; only three types of molecules can be produced: linear, cyclic, and branched. The cyclic molecules cannot contain branch units. Hence regardless of the content of RA_n, cross-linked molecules cannot be produced. This is the same result considered in vinyl-type branching. On the other hand, in the system AB + RA_mB_n where A can only react with B and vice versa and $m + n \geqslant 3$, then in addition to linear, cyclic and branched molecules, cross-linked material may result.

Divinyl compounds are tetrafunctional. Monomers such as butadiene can be polymerized to high molecular weight, soluble materials because the "internal" double bonds of 1,4 polymerization or the pendant double bond of a 1,2 polymerization are less reactive than the double bonds of the monomer. During vulcanization, some of the remaining double bonds react with the vulcanizer molecule to form the cross-linked polymer. On the other hand, monomers such as divinyl benzene cannot be homopolymerized to high molecular weight soluble materials because of the easc of ring formation. High molecular weight branched and partially cross-linked addition polymers can be made by adding small amounts of divinyl benzene to the monomer mix.

The distribution functions in nonlinear systems generally omit ring formation, hence they are not applicable near the gel point. The gel point occurs when molecules of infinite molecular weight appear. Because many molecules are present, the number-average molecular weight may still be finite and only slowly increasing at the conversion where the weight-average molecular weight approaches infinity. In general, the critical conversion, α_c, is given by

$$\alpha_c = 1/(\bar{f} - 1) \tag{6}$$

where $\bar{f}$ is the average functionality of the system.

In the sections that follow, we consider grafting and branching reactions in vinyl polymerization, the Stockmayer distribution function for condensation polymers when ring formation is excluded, some specific distributions of nonlinear condensation systems, ring formation, and gel points.

2. Vinyl Polymerization. Self-Grafting

Constant rate of initiation, monomer concentration invariant, transfer to monomer and to polymer, termination by second-order disproportionation.

Bamford and Tompa (1) considered the problem of branch formation by means of the transfer-to-polymer reaction:

$$R_n^* + P_s \rightarrow P_n + R_s^* \qquad k_{tr,p} \tag{1}$$

The velocity of this reaction is assumed to be proportional to the number of monomer units in the polymer molecule under attack; in this case

$$dR_s^*/dt = k_{tr,p} s P_s R_n^* \tag{2}$$

The assumption is made that the monomer and initiator concentrations are maintained constant by adding monomer and initiator in suitable amounts. Thus, the various molecular species are not invariant with time, as is usually assumed. By LaPlace transforms and a reduced time variable, ξ, where

$$\xi = k_{tr,p}(R_I/2k_{t,d})^{\frac{1}{2}} \bar{r}_n t \tag{3}$$

the differential equations can be solved. The distribution for all molecules, branched and unbranched, is

$$W(r) = (1/\bar{r}_n) \exp[-r(1 + \xi)/\bar{r}_n][(2 + \xi)/\xi]^{\frac{1}{2}} \tilde{I}_1(y) \tag{4}$$

(Bamford and Tompa, 1) where $\tilde{I}_1(y)$ is the modified Bessel function of the first kind of order one and

$$y = r\xi^{\frac{1}{2}}(2 + \xi)^{\frac{1}{2}}/\bar{r}_n \tag{5}$$

The average degrees of polymerization are

$$\bar{r}_n = k_p M/(k_{tr,m} M + \{2R_I k_{t,d}\}^{\frac{1}{2}}) \tag{6}$$

$$\bar{r}_w = \bar{r}_n (2 + \xi) \tag{7}$$

The distribution for unbranched molecules is

$$W(r, 0) = (r/\bar{r}_n^{\,2})(1 + \xi/2) \exp[-r(1 + \xi)/\bar{r}_n] \tag{8}$$

(Bamford and Tompa, 1) and that for molecules containing b branches,

$$W(r, b) = (r/\bar{r}_n^{\,2}) \exp[-r(1 + \xi)/\bar{r}_n]$$

$$\times [(2 + \xi)/2b!(b + 1)!](y/2)^{2b} \tag{9}$$

(Bamford and Tompa, 1).

The maximum of these distributions occurs at the respective number-average degrees of polymerization $\bar{r}_n|_b$. The number-average and weight-average degrees of polymerization for the individual distributions are given by

$$\bar{r}_n|_b = \bar{r}_n (2b + 1)/(1 + \xi) \tag{10}$$

$$\bar{r}_w|_b = r_n (2b + 2)/(1 + \xi) \tag{11}$$

The average number of branch points per molecule of size

$\underline{r}$ is given by

$$\bar{b}_n(r) = y\tilde{I}_2(y)/2\tilde{I}_1(y) \tag{12}$$

where $\tilde{I}_2(y)$ is the modified Bessel function of the first kind of order two.

The number-average number of branch points per molecule is

$$\bar{b}_n = \xi/2 \tag{13}$$

The distribution function (4) for the whole polymer is given in Figure 5.2.1 for various values of the parameter ξ. Several of the branched distribution functions W(r,b) are given in Figure 5.2.2 for $\xi = 2$. As the number of branches per molecule becomes large, the distribution functions overlap and become rather indistinguishable in a graph such as Figure 5.2.2. To show the individual curves more clearly, the fraction of molecules of size $\underline{r}$, with $\underline{b}$ branches, relative to the entire distribution, is shown in Figure 5.2.3.

Bamford and Tompa (1) also consider the problem of grafting onto a preformed polymer whose initial distribution is either (i) of constant length $\bar{r}_n$ or (ii) a geometric distribution. Further details are given in their paper.

Curve	a	b	c	d
ξ	0	0.2	0.67	2.0

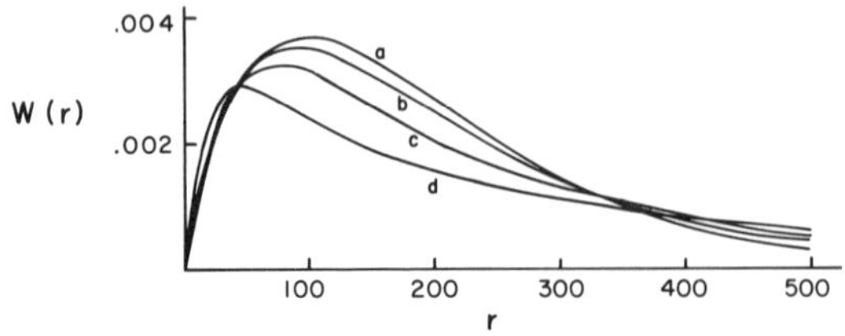

Figure 5.2.1 - Weight fraction distribution as a function of $\underline{r}$ for several values of the parameter ξ for constant rate of initiation, monomer concentration invariant, transfer to monomer and to polymer, termination by disproportionation, $\bar{r}_n = 100$. (After Bamford and Tompa, 1).

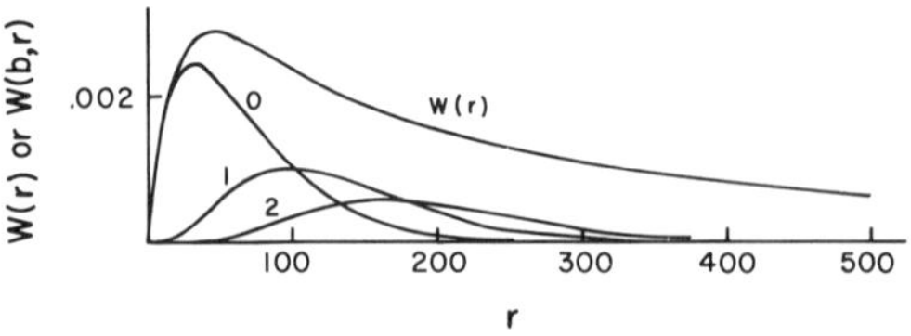

Figure 5.2.2 - Weight fraction distribution as a function of $\underline{r}$ for the whole polymer and molecules which contain zero one, and two branches per molecule for $\xi = 2$. Same conditions as in Figure 5.2.1.

3. <u>Vinyl Polymerization. Calculation of the Moments of the Distribution. Terminal Double Bond Polymerization</u>

Constant rate of initiation, monomer concentration varies, transfer to monomer and to polymer, terminal double bond polymerization, termination predominantly by transfer

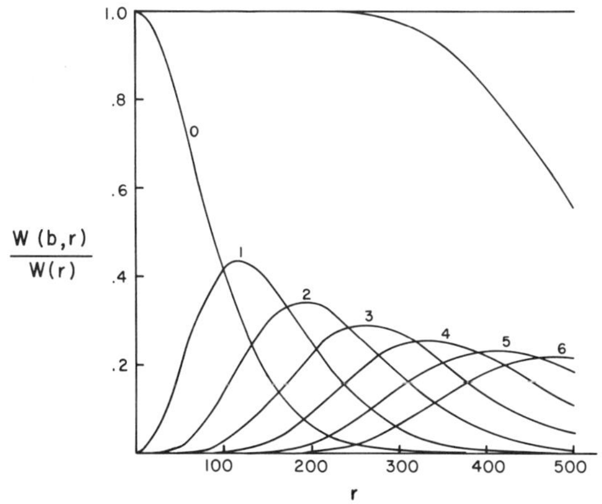

Figure 5.2.3 - The relative weight fraction distribution as a function of r for nonbranched and branched molecules when $\xi = 2$. The curve in the upper right corner gives $\sum_{b=0}^{b} W(b,r)/W(r)$. Same conditions as in Figure 5.2.1.

Bamford and Tompa (1) give a general method for finding the moments of a distribution as a function of conversion. The method is particularly useful when analytical solutions to the distribution equation cannot be found. From the kinetic scheme, one writes the differential equation for inactive polymer molecules of size r, dP_r/dt, and for active polymer molecules of size r, dR_r^*/dt. The following definitions are made:

$$Q_n = \sum_r r^n P_r \tag{1}$$

$$Y_n = \sum_r r^n R_r^* \tag{2}$$

Here Q_0 is the total number of inactive polymer molecules and Q_1 is the total number of polymerized monomer molecules in the inactive polymer molecules. For high molecular weight polymers, the Taylor series expansion of $\sum_{r=1}^{\infty} r^n P_{r-1} = \sum_{s=0}^{\infty} (s + 1)^n P_s$ may be discontinued after the second term to give the approximation

$$\sum_{r=1}^{\infty} r^n P_{r-1} = \sum_{r=1}^{\infty} r^n P_r + n \sum_{r=1}^{\infty} r^{n-1} P_r \tag{3}$$

The first three moments of the distribution may be obtained by substituting n = 0, 1, or 2 into the differential equations for $dQ_n/dt = 0$, and $dY_n/dt = 0$ and eliminating Y_1 and Y_2. Graessley, Mittelhauser, and Maramba (2) used this method to find $\bar{r}_n$ and $\bar{r}_w$ for the case when so little initiator is added to the system that only a small fraction of polymer chains are formed by mutual termination, the major fraction being formed by the transfer-to-monomer reaction. In their scheme, each act of transfer causes a terminal double bond to be incorporated into a chain which may later participate in polymerization thereby placing a branch in the chain:

$$\bar{r}_n = \frac{1 - K}{C_m} \left[\frac{c}{(1 - c)^K - (1 - c)} \right], \qquad K \neq 1 \tag{4}$$

$$= -c/[C_m(1 - c) \ln (1 - c)], \qquad K = 1 \tag{5}$$

where

$$C_m = k_{tr,m}/k_p$$

$$c = (M_o - M)/M_o \tag{6}$$

$$K = k_{p,p}/k_p$$

where $\underline{K}$ is ratio of the propagation reactivity constant of the terminal double bond to that of the monomer.

For the general case of transfer to monomer, to polymer, and terminal double bond polymerization, let $\bar{r}_w = Q/c$; then

$$\frac{dQ}{dc} = \frac{2\left[1 + \frac{Kc}{1 - c}\right]\left[1 + \frac{Kc}{1 - c} + \frac{C_p Q}{1 - c}\right]}{[C_m + C_p\ c/(1 - c)]} \tag{7}$$

where $C_p = k_{tr,p}/k_p$. Equation (7) can be integrated numerically.

If terminal double bond polymerization does not occur, then K = 0; letting $C_p/C_m = a$, then

$$\bar{r}_n = 1/C_m \tag{8}$$

$$\bar{r}_w = \frac{1}{C_m}\left\{\frac{[1 + (a - 1)c]^{[2a/(a-1)]} + 2c - 1}{c(a + 1)}\right\} \qquad a \neq 1 \tag{9}$$

$$= \frac{1}{C_m}\left\{\frac{\exp\ (2c) + 2c - 1}{2c}\right\} \qquad a = 1 \tag{10}$$

If $C_p = 0$ and $\underline{K}$ is finite, then

$$\bar{r}_w = \frac{2}{C_m} \{(1 - 2K + 2K^2) + K^2 c/(1 - c)$$

$$- [2K(1 - K)/c] \ln (1 - c)\} \tag{11}$$

The parameter $\bar{r}_n C_m$ is given in Figure 5.3.1 for equation (4) as a function of conversion, $\underline{c}$, for various values of $\underline{K}$; $\bar{r}_n$ approaches infinity as $\underline{c}$ approaches unity because each polymer molecule formed contains a terminal double bond which can react. The initial slope of $\bar{r}_n$ versus conversion is positive rather than negative as shown in Figure 2.3.3, page 73. Recalling equation (2.4.12),

$$\bar{r}_n = k_p(M_o - M)/\Big\{[\{2R_I\ k_{t,d}\}^{\frac{1}{2}} + k_{tr,s}S] \ln M_o/M$$

$$+ k_{tr,m}(M_o - M)\Big\}$$

By setting both $k_{t,d}$ and $k_{tr,s}$ equal to zero, we see that $\bar{r}_n = 1/C_m$. Thus the negative slope in Figure 2.3.3 arises from transfer to solvent and termination reactions. These reactions are excluded from the present kinetic schemes. Hence molecules are formed predominantly by transfer to monomer. The number-average molecular weight is independent of the extent of the transfer-to-polymer reaction. This follows because the act of transfer does not change the number of polymer molecules present. The same is not true with respect to the molecular weight distribution as measured by $\bar{r}_w/\bar{r}_n$; the transfer-to-polymer reaction causes $\bar{r}_w/\bar{r}_n$ to increase (compare Figure 5.3.2), but remains finite at large conversion. In a batch-type polymerization, the transfer-to-polymer reaction by itself cannot lead to gelation. On the other hand, terminal bond polymerization causes $\bar{r}_w$ to increase faster than $\bar{r}_n$, compare Figure 5.3.3, forming gel at high conversion. In a continuous-flow, stirred-tank reactor, the concentration of each species is independent of time after equilibrium has been reached, usually 5 or 6 dwell times. Although the average lifetime of a molecule in the reactor is given by the dwell time, θ, it is possible for an individual molecule to remain in the reactor for considerably longer periods of time. Hence if the transfer-to-polymer reaction occurs, some molecules can become exceedingly large with many branches; in contrast to batch-type polymerization (Figure 5.3.2), infinite weight-average molecular weights can occur at intermediate conversions (Figure 5.3.4). The conversion in a continuous-flow, stirred-tank reactor is given by

$$c = [(M)_{in} - (M)_{out}]/(M)_{in}$$

$$= \text{constant for any given run when } t > 6\theta \qquad (12)$$

$$c = \frac{(k_p + k_{tr,m})\, R_I \theta^2}{1 + (k_p + k_{tr,m})\, R_I \theta^2} \qquad (13)$$

Equation (13) is derived on the basis that termination

Curve	a	b	c	d	e
K	0.1	0.25	0.5	1.0	2.0

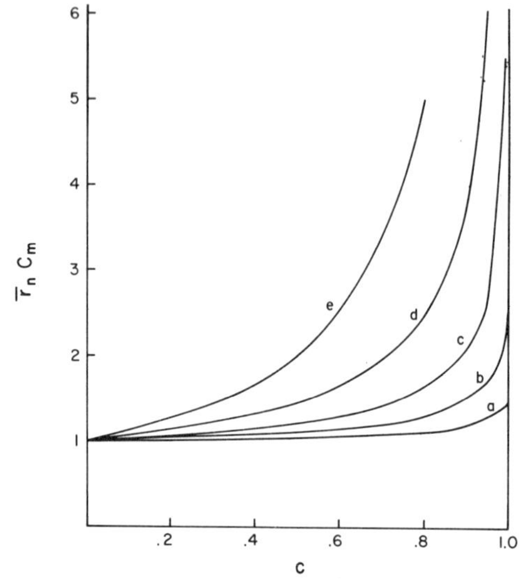

Figure 5.3.1 - Effect of terminal bond polymerization on $C_n\bar{r}_n$ as a function of conversion parameter $K = k_{p,p}/k_p$. (After Graessley et al., 2).

reactions do not occur. If, on the other hand, $2k_tR^{*2} \gg R^*/\theta$, that is, if termination reactions are important and high molecular weight polymer is produced, then

$$c = \frac{k_p(R_I/2k_t)^{\frac{1}{2}}\,\theta}{1 + k_p(R_I/2k_t)^{\frac{1}{2}}\,\theta} \tag{14}$$

The average degrees of polymerization in a continuous-flow, stirred-tankreactor with termination predominantly by transfer to monomer at steady state conditions are

Curve	a	b	c	d	e
C_p/C_m	.1	.25	.5	1	2

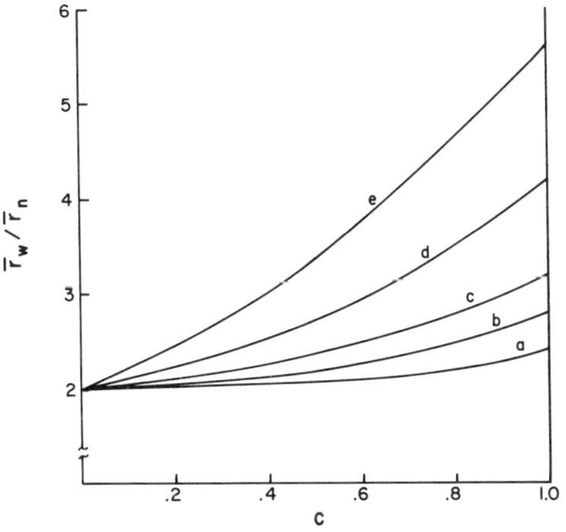

Figure 5.3.2 - Effect of transfer to polymer on the dispersion ratio as a function of conversion and the parameter, C_p/C_m, the ratio of the transfer constants to polymer and to monomer, $\bar{r}_n = 1/C_m$ for all conversions. Batch polymerization. (After Graessley et al., 2).

$$\bar{r}_n = 1/C_m \tag{15}$$

$$\bar{r}_w = 2/\{C_m[1 - ac/(1 - c)]\} \tag{16}$$

(Graessley, 2b).

4. <u>Vinyl Polymerization</u>. <u>Long Chain Branching</u>

Constant rate of initiation, monomer concentration invariant, transfer to monomer and to polymer, termination by disproportionation.

Curve	a	b	c	d	e
$k_{p,p}/k_p$	0.10	0.25	0.5	1.0	2.0

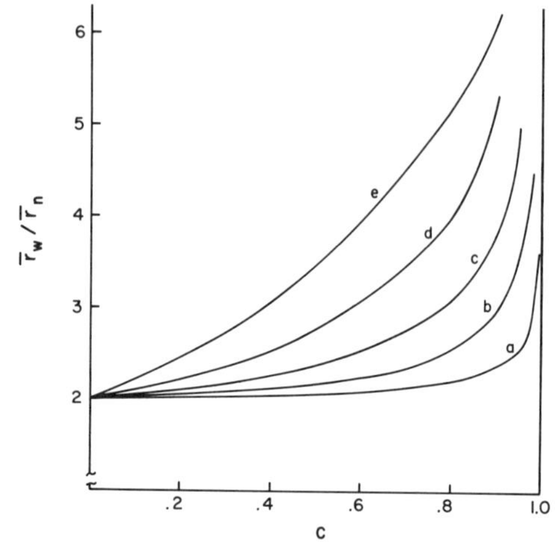

Figure 5.3.3 - Effect of terminal bond polymerization on the dispersion ratio as a function of conversion and the parameter $k_{p,p}/k_p$. Batch polymerization. (After Graessley et al., 2).

Beasley also considered the case of long chain branching in a continuous-flow, stirred-tank reactor. Let β be the fraction of monomers in branches. The fraction of monomers in "chains," that is, not in branches, is given by $(1 - \beta)$. Let $1/a$ be the average number of monomers in chains and

$$1/a = \bar{r}_n(1 - \beta) \tag{1}$$

$$\beta = k_{tr,p}P/k_p Ma \tag{2}$$

Curve	a	b	c	d	e
a	.1	.25	.5	1.0	2.0

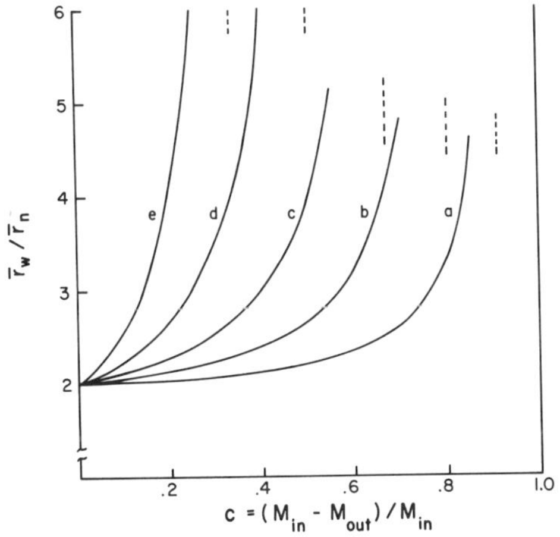

Figure 5.3.4 - Effect of the transfer-to-polymer reaction on the dispersion ratio as a function of "conversion" in a continuous-flow, stirred-tank reactor at steady state. The monomer concentration in the tank is invariant with time; hence conversion is a function of polymerization conditions. The dashed vertical lines give the values of conversion at which $\bar{r}_w = \infty$. $\bar{r}_n = 1/C_m$ for all conversions. $a = C_p/C_m$. (After Graessley, 2b).

The rate of the transfer-to-polymer reaction is dependent upon the number of units in the polymer; see equation (5.1.1).

Beasley's treatment of the problem differs from Bamford and Tompa's primarily in the use of a continuous-flow, stirred-tank reactor in which all molecular species essentially remain invariant with time. As pointed out in the preceding section, polymers of infinite molecular weight can be formed in a reactor of this type because some molecules can remain in the reactor for long periods of time.

The distribution function for molecules of size $\underline{r}$ containing $\underline{b}$ branches is

$$W(b, r)$$

$$= \frac{(1 - \beta)(a^2 r)[ar - (1/\beta)\ \ln(1 + \beta ar)]^b\ \exp(-ar)}{b!(1 + \beta ar)} \quad (3)$$

(Beasley, 3). The fraction of unbranched molecules is

$$\sum_{r=0}^{\infty} F\ (0, r) = (1/\beta)\ \exp(1/\beta) \int_{1/\beta}^{\infty} (1/s)\ \exp(-s)\ ds \quad (4)$$

where the integral is the exponential integral -Ei(-x), reference 8, Chapter 2. The entire distribution is

$$W(r) = \sum_{b=0}^{\infty} W(b, r) = (1 - \beta)(a^2 r)/[1 + \beta ar]^{(1+1/\beta)} \quad (5)$$

(Beasley, 3). The number-average number of branches of molecules of size $\underline{r}$ is

$$\bar{b}_n(r) = ar - (1/\beta)\ \ln(1 + \beta ar) \quad (6)$$

The average molecular weights are

$$\bar{r}_n = 1/a(1 - \beta) \qquad \beta < 1 \quad (7)$$

$$\bar{r}_w = 2/a(1 - 2\beta) \qquad \beta < \tfrac{1}{2} \tag{8}$$

$$\bar{r}_z = 3/a(1 - 3\beta) \qquad \beta < 1/3 \tag{9}$$

$$\bar{b}_n = [1/(1 - \beta)] - 1 \qquad \beta < 1 \tag{10}$$

$$\bar{b}_w = [1/(1 - 2\beta)(1 - \beta)] - 1 \qquad \beta < \tfrac{1}{2} \tag{11}$$

The distribution (3) is given in Figure 5.4.1 for various values of β. At $\beta = \frac{1}{2}$, $\bar{r}_w = \infty$ [compare equation (8)]; however, the lower molecular weight species in the distribution still have a finite and determinable distribution because $\bar{r}_n$ remains finite. Figure 5.4.2 shows an exploded view of the high molecular weight tails. The tail for the distribution $\beta = \frac{1}{2}$ is observed to approach the r axis at a slowly decreasing rate. The distribution of the branched molecules in the $\beta = \frac{1}{2}$ polymer is given in Figure 5.4.3.

Curve	a	b	c	d
β	0	0.1	0.25	0.50
$\bar{r}_w$	200	225	300	∞

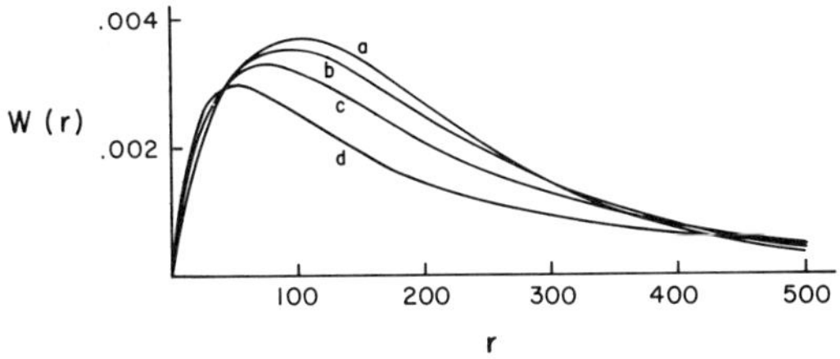

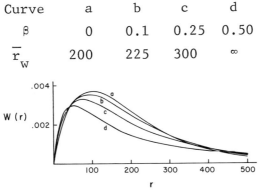

Figure 5.4.1 - Weight fraction distribution as a function of r and the parameter β, the fraction of monomers in branches, for constant rate of initiation, transfer to

monomer and to polymer, termination by disproportionation, constant flow reactor, $\bar{r}_n = 100$. (After Beasley, 3).

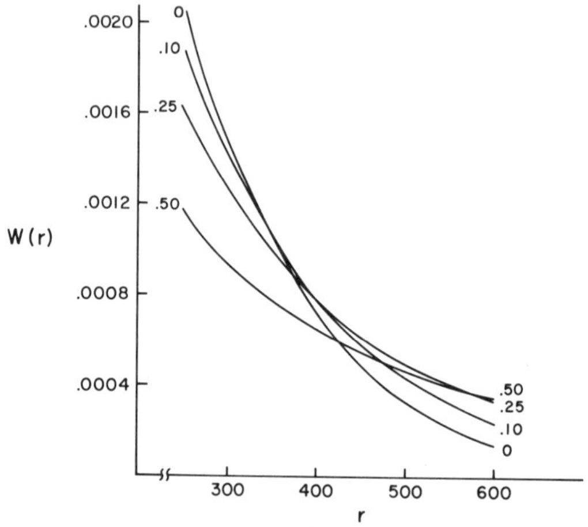

Figure 5.4.2 - Exploded view of the high molecular weight tail of Figure 5.4.1.

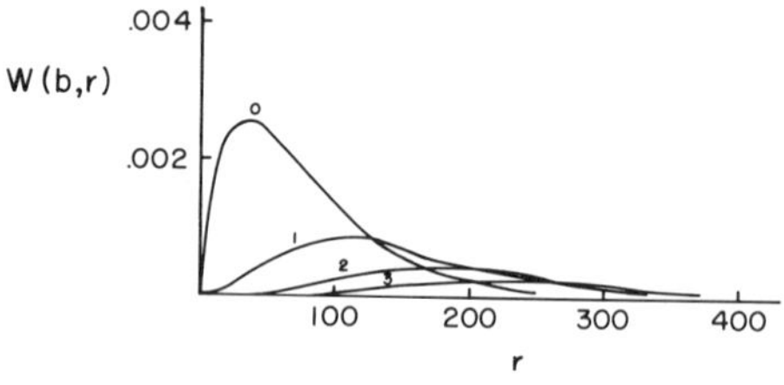

Figure 5.4.3 - Weight fraction distribution as a function of r for molecules containing b branches when $\bar{r}_n = 100$, $\beta = \frac{1}{2}$ (after Beasley, 3).

At b values of 4 or greater, the curves blend together; the fraction of the distribution W(b, r)/W(r) is given in Figure 5.4.4. The average number of branches per molecule is given in Figure 5.4.5. At a number average of one branch per molecule, $\beta = \frac{1}{2}$, $\bar{b}_w$ and $\bar{r}_w$ become infinite.

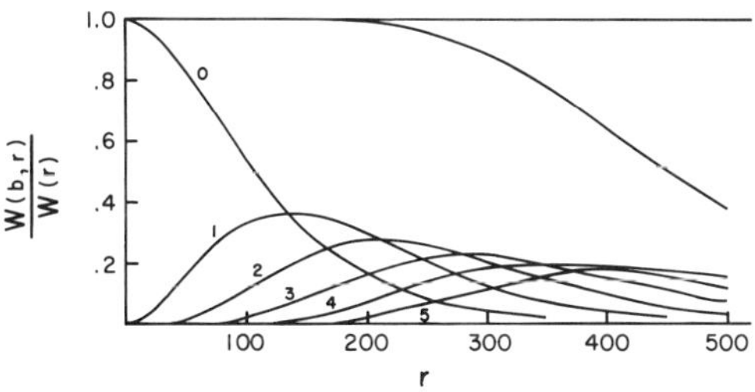

Figure 5.4.4 - The relative weight fraction distribution as a function of r for non-branched and branched molecules when $\beta = \frac{1}{2}$, $\bar{r}_n = 100$. (After Beasley, 3). The curve in the upper right hand corner gives $\sum_{b=0}^{5} W(b,r)/W(r)$.

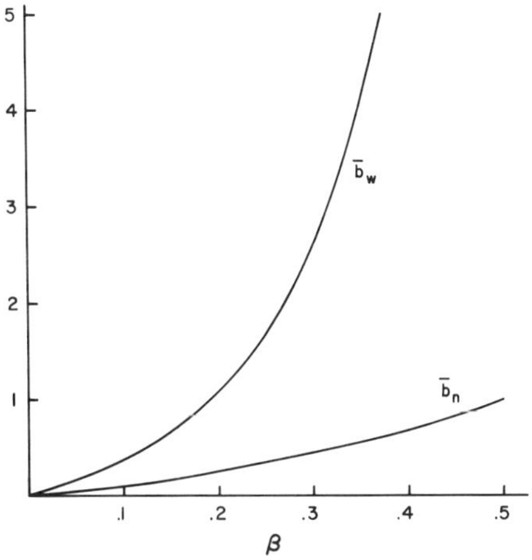

Figure 5.4.5 - The average number of branches per molecule as a function of the parameter β. (Courtesy *J. Am.*

Chem. Soc.) (Beasley, 3).

5. Vinyl Polymerization. Branching Density

Branching density as a function of conversion. Branches formed by polymerization through a vinyl group (diene polymers) or by transfer-to-polymer reaction.

The branching density, ρ, is given by the ratio of the number of branches to the number of monomers polymerized:

$$\rho = -(k_{tr,p}/k_p)\{1 + [M_o/(M_o - M)] \ln(M/M_o)\} \qquad (1)$$

$$= -s(k_{p,p}/k_p)\{1 + [M_o/(M_o - M)] \ln(M_o/M_o)\} \qquad (2)$$

(Flory, 4) where $k_{tr,p}$ is the rate constant for transfer to polymer and $k_{p,p}$ is the rate constant for propagation through a double bond in the polymer. If the double bond is at the end of the polymer, a "three-armed" branch point is formed, similar to that formed by the transfer-to-polymer reaction, and $s = 1$. If, however, the double bond is an internal one, a "four-armed" branch point is formed and $s = 2$.

In a continuous-flow, stirred-tank reactor, the branching density is given by

$$\rho = (k_{tr,p}/k_p)[(M_o - M)/M] \qquad (3)$$

(Graessley, 2b).

6. The Stockmayer Distribution Function. Formation of rings excluded prior to gelation.

Stockmayer (5) presents a generalized distribution formula for a variety of monomers containing end groups of type A which can only react with a variety of monomers containing end groups of type B. In the original mixture there are A_1, A_2, A_3, ..., A_i, ... moles of reactants bearing respectively f_1, f_2, f_3, ..., f_i, ... functional groups of type A each, together with B_1, B_2, ..., B_i, ... moles of reactants of functionalities g_1, g_2, ..., g_i, ... of groups of type B. All functional groups of a given type are equally reactive and ring formation does not occur appreciably, which obviously is not true near the gel point. The system reacts until a fraction α of the A groups and a fraction β of the B groups have reacted. Furthermore,

$$\alpha \Sigma_i f_i A_i = \beta \Sigma_j g_j B_j$$

Now $N\{m_i, n_j\}$ represents the number of moles of that species which consists of m_1, m_2, ..., m_i, ... monomer units of the A type combined with n_1, n_2, ..., n_j, ... units of the B type.

$$N\{m_i, n_j\} = [K(\Sigma_i f_i m_i - \Sigma_i m_i)!\ (\Sigma_j g_j n_j - \Sigma_j n_j)!$$

$$\times \prod_i (x_i^{m_i}/m_i!) \prod_j (y_j^{n_j}/n_j!)]/[(\Sigma_i f_i m_i - \Sigma_i m_i$$

$$- \Sigma_j n_j + 1)! \, (\Sigma_j g_j n_j - \Sigma_j n_j - \Sigma_i m_i + 1)!] \tag{1}$$

(Stockmayer, 5) where

$$x_i = \frac{f_i A_i \beta (1 - \alpha)^{f_i - 1}}{(\Sigma_i f_i A_i)(1 - \beta)} \tag{2}$$

$$y_j = \frac{g_j B_j \alpha (1 - \beta)^{g_j - 1}}{(\Sigma_j g_j B_j)(1 - \alpha)} \tag{3}$$

$$K = (\Sigma_i f_i A_i)(1 - \alpha)(1 - \beta)/\beta = (\Sigma_j g_j B_j)(1 - \alpha)(1 - \beta)/\alpha \tag{4}$$

an example will illustrate the use of Equation (1). Suppose that we have the monomers acetic acid (CH_3COOH, A_1, $f_1 = 1$) and adipic acid ($HOOC(CH_2)_4COOH$, A_2, $f_2 = 2$) reacting with ethylene glycol ($HOCH_2CH_2OH$, B_2, $g_2 = 2$) and glycerol ($HOCH_2CHOHCH_2OH$, B_3, $g_3 = 3$) (all hydroxyls of the glycerol are considered equally reactive). What is the number of molecules which contain exactly 1 acetic acid unit, 4 adipic acid units, 3 glycol units, and 2 glycerol units? It is N(1, 4, 3, 2) and

$$N(1, 4, 3, 2) = \frac{K(1 + 8 - 1 - 4)!}{(1 + 8 - 1 - 4 - 3 - 2 + 1)!}$$

$$\times \frac{(6 + 6 - 3 - 2)!}{(6 + 6 - 3 - 2 - 1 - 4 + 1)!} \left(\frac{x_1^1}{1!}\right)\left(\frac{x_2^4}{4!}\right)\left(\frac{y_2^3}{3!}\right)\left(\frac{y_3^2}{2!}\right) \tag{5}$$

$$x_1 = \frac{A_1 \beta}{(A_1 + 2A_2)(1 - \beta)} \qquad x_2 = \frac{2A_2 \beta(1 - \alpha)}{(A_1 + 2A_2)(1 - \beta)}$$

$$y_2 = \frac{2B_2 \alpha(1 - \beta)}{(2B_2 + 3B_3)(1 - \alpha)} \qquad y_3 = \frac{3B_3 \alpha(1 - \beta)^2}{(2B_2 + 3B_3)(1 - \alpha)}$$

$$K = (A_1 + 2A_2)(1 - \alpha)(1 - \beta)/\beta \tag{6}$$

If each species $\underline{i}$ has an effective molecular weight, M_i, which is lower than the original molecular weight by the term $W_o f_i/2$, where W_o is the molecular weight of the by-product, then the weight of molecules present is

$$W = \Sigma_i M_i A_i + \Sigma_j B_j M_j \tag{7}$$

The number of molecules of the end of the reaction is, neglecting by-product,

$$N = \Sigma_i A_i + \Sigma_j B_j - \alpha \Sigma_i f_i A_i \tag{8}$$

The number-average molecular weight is

$$\overline{M}_n = W/N \tag{9}$$

The weight-average molecular weight is

$$\overline{M}_w = \left\{ \beta \frac{\Sigma_i M_i^2 A_i}{\Sigma_i f_i A_i} + \alpha \frac{\Sigma_j M_j^2 B_j}{\Sigma_j g_j B_j} \right.$$

$$+ \frac{\alpha\beta[\alpha(f_e - 1)M_b^2 + \beta(g_e - 1)M_a^2 + 2M_aM_b]}{1 - \alpha\beta(f_e - 1)(g_e - 1)}\Bigg\}$$

$$\div \left\{\beta\frac{\Sigma_i M_i A_i}{\Sigma_i f_i A_i} + \alpha\frac{\Sigma_j M_j B_j}{\Sigma_j g_j B_j}\right\} \tag{10}$$

where

$$f_e = (\Sigma_i f_i^2 A_i)/(\Sigma_i f_i A_i)$$

$$g_e = (\Sigma_j g_j^2 B_j)/(\Sigma_j g_j B_j)$$

$$M_a = (\Sigma_i M_i f_i A_i)/(\Sigma_i f_i A_i)$$

$$M_b = (\Sigma_j M_j g_j B_j)/(\Sigma_j g_j B_j) \tag{11}$$

The gel point is

$$(\alpha\beta)_c = 1/(f_e - 1)(g_e - 1) \tag{12}$$

The system comprizing AA plus BB molecules plus a terminator A and a branching agent A_4 has been studied by Nakamura, Yokouchi, Ito, Miura, and Fuyii (6). They present the mole fraction of molecules of various structures as a function of A_1, A_4, and conversion. Figure 5.6.1 shows how linear and branched molecules vary with terminator concentration in the presence and absence of A_4 branching agent, for 99% conversion.

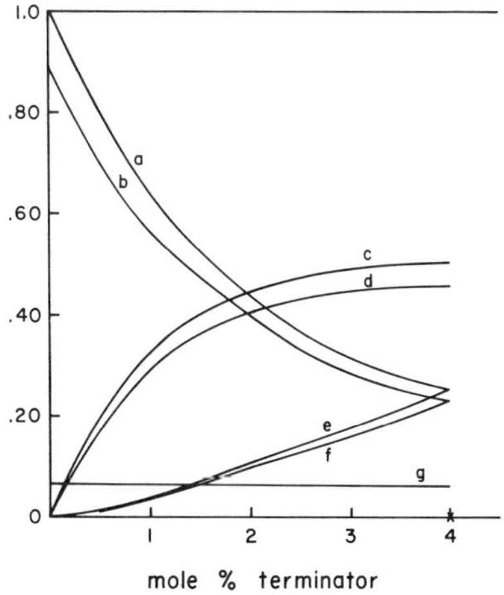

Figure 5.6.1 - Fraction of linear molecules which contain none (a, b) one (c, d) or two (e, f) terminator molecules as a function of terminator concentration at 99% reaction, equimolar amounts of reactants AA and BB, zero (a, c, e) or 0.5 mol % (b, d, f) of tetrafunctional branch units. The fraction of molecules containing only one tetrafunctional unit, (g) varies from 6.81 to 5.99% as mol % terminator varies from 0 - 4%. The fraction of molecules containing one tetrafunctional unit and four terminator units is 0.36% at 4% terminator concentration (x) (after Nakamura et al., 6).

7. Cross-linking or Coupling of a Polymer with a Known Primary Distribution. Formation of rings excluded prior to gelation (4, 7-9).

Some coupling reactions are considered in Section 4.7 for linear systems. If each molecule of the known primary distribution has the same end group, then the number distribution can be computed with the aid of Section 5.6, by letting the various quantities m_i represent the integral number of molecules of size _i_ in the primary distribution, the n_i to represent the number of molecules of cross-linking agent.

Alternately, let us consider the distribution function for a polymer before and after a certain extent of cross-linking has occurred. To distinguish between the molecules of the primary distribution before cross-linking and the molecules after cross-linking, we call the former _chains_, which need not imply linearity to the chains of the primary distribution, and we use the word _molecules_ to mean molecules after cross-linking which are made up of chains. Thus there are N_1 chains containing one monomer, N_2 chains containing 2 monomers, N_s chains containing _s_ monomers, and so on. The number-average degree of polymerization of the primary distribution is

$$\bar{r}_n^{\,o} = \Sigma s N_s / \Sigma N_s \tag{1}$$

while the weight-average degree of polymerization of the primary distribution is

$$\bar{r}_w^{\,o} = \Sigma s^2 N_s / \Sigma s N_s = \Sigma s w_s \tag{2}$$

where w_s is the weight fraction of chains containing _s_ monomers

$$w_s = s N_s / \Sigma s N_s \tag{3}$$

Each monomer unit in a chain contains, on the average, ρ cross-linking sites. The cross-linking sites must be randomly distributed among the ΣsN_s monomer units. We now permit a fraction α of the cross-linking sites to react but require that rings do not form prior to gelation. (We have a Maxwell demon in the reactor who prevents ring formation reactions!) The quantity $\gamma \equiv \alpha\rho\bar{r}_n^{\,o}$ is the number-average number of cross-links per chain; it is also known as the cross-linking index. The gel point given by

$$\alpha_c = 1/\rho(\bar{r}_w^{\,o} - 1) \tag{4}$$

hence

$$\gamma_c \approx \bar{r}_n^{\,o}/\bar{r}_w^{\,o} \tag{5}$$

After cross-linking we have molecules made up from the original chains. One new molecule selected at random may be composed of n_1 chains of size 1, n_2 chains of size 2, n_3 chains of size 3, ... n_s chains of size $\underline{s}$, and so on. This large set of numbers with a variety of subscripts is abbreviated by the symbol $\{n_s\}$. Thus $m\{n_s\}$ will give the number of molecules described by the particular set $\{n_s\}$. It is particularly important to realize that no distinction is made as to the spatial arrangement of the n_1, n_2, n_3, ..., n_s chains in the set $\{n_s\}$. If we set $\rho = 1$, the $m\{n_s\}$ is given by

$$m\{n_s\} = (\Sigma sN_s)\frac{(\Sigma sn_s - \Sigma n_s)!}{(\Sigma sn_s - 2\Sigma n_s + 2)!}\alpha^{\Sigma n_s - 1}$$

$$\times (1 - \alpha)^{(\Sigma sn_s - 2\Sigma n_s + 2)} \prod_s (w_s^{n_s}/n_s!) \tag{6}$$

As an example, consider a molecule in which $n_1 = 3$, $n_2 = 5$, $n_3 = 4$, and $n_4 = 1$, then

ΣsN_s is the total number of monomers in the system

$\Sigma sn_s = 1 \cdot 3 + 2 \cdot 5 + 3 \cdot 4 + 4 \cdot 1 = 29$ = the number of monomers in this particular molecule

$\Sigma n_s = 13$ = the number of chains in the molecule

$\Sigma sn_s - 2\Sigma n_s + 2 = 5$

$$m\{n_s\} = \frac{(\Sigma sN_s)(16)!}{5!}\alpha^{12}(1 - \alpha)^5\left(\frac{w_1^{\,3}}{3!}\right)\left(\frac{w_2^{\,5}}{5!}\right)\left(\frac{w_3^{\,4}}{4!}\right)\left(\frac{w_4^{\,1}}{1!}\right)$$

where w_s is given by equation (3). The weight-average degree of polymerization after cross-linking is

$$\bar{r}_w = \bar{r}_w^{\,o}(1 + \alpha)/[1 - \alpha(\bar{r}_w^{\,o} - 1)] \tag{7}$$

and again with $\rho = 1$,

$$\alpha_c = 1/(\bar{r}_w^{\,o} - 1) \tag{8}$$

When $\rho < 1$, then $\gamma = \alpha\rho\bar{r}_n^{\,o}$, $\gamma_c = \bar{r}_n^{\,o}/\bar{r}_w^{\,o}$

$$\bar{r}_n = \bar{r}_n^{\,o}/(1 - \gamma/2) \tag{9}$$

$$\bar{r}_w = \bar{r}_w^{\,o}/(1 - \gamma/\gamma_c) \qquad (10)$$

If all the primary chains have the same size s (the uniform distribution), equation (5) reduces to

$$m(n) = sN\alpha^{n-1}(1 - \alpha)^{(sn-2n+2)}\left[\frac{(sn - n)!}{n!(sn - 2n + 2)!}\right] \qquad (11)$$

(Stockmayer, 7) where n is the number of primary molecules in the cross-linked molecule. By use of Stirling's formula and assuming $s \gg 1$, we obtain

$$W(n) = n^{n-1}(\gamma e^{-\gamma})^n/\gamma n! \qquad (12)$$

(Flory, 9, Stockmayer, 7). Table 5.7.1 contains a number of values of W(n) for $0.01 \leqslant \gamma \leqslant 0.99$.

The fraction of molecules of size r containing i cross-linked units, provided that $i \leqslant r$, is

$$F_r(i) = [r!/(r - i)!i!]\,\rho^i(1 - \rho)^{r-i} \qquad (13)$$

(Flory, 9).

The weight fraction of soluble material is γ'/γ, where γ' is the lower root of

$$\beta = \gamma\exp(1 - \gamma) \qquad (14)$$

Amemiya (10) has derived similar equations for the uniform distribution as well as for the Poisson distribution.

TABLE 5.7.1

W(n) for Cross-linked Uniform Distribution

	Cross-linking Index						
γ	0.01	0.10	0.25	0.50	0.75	0.90	0.99
n	Number of Primary Molecules						
1	0.990	0.905	0.779	0.606	0.473	0.406	0.372
2	0.010	0.082	0.153	0.184	0.168	0.149	0.128
3		0.011	0.044	0.084	0.089	0.082	0.076
4		0.002	0.015	0.045	0.056	0.053	0.049
5			0.006	0.027	0.039	0.038	0.035
6			0.002	0.017	0.028	0.029	0.027
7			0.001	0.011	0.022	0.023	0.022
8				0.007	0.017	0.018	0.018
9				0.005	0.014	0.016	0.015
10				0.004	0.011	0.013	0.013
$\sum_{n=1}^{10} W(n)$	1.000	1.000	1.000	0.990	0.917	0.847	0.755

8. Homopolymer of ARB_{f-1}. Where A can react only with B. The B's may have different reactivities. Formation of rings excluded prior to gelation.

In a polymer of this type, one end of a given molecule can react with the opposite end to form a closed ring. If $f \geq 3$, a highly branched structure results at low conversions. In the strictest sense, such structures are not three-dimensional gels, as only one ring can form per molecule, but infinitely large molecules can be formed at intermediate conversions. The theory of highly branched, unidirectional structures suffers from the same inadequacy as the theory of gel formation because the principle of equal reactivity is no longer valid when large molecules are formed. The statistical probability of any given group A being able to react with a group B will depend on the ability of A and B to get close enough to each other in order to react. They may be sterically prevented from attaining the requisite proximity. One means of avoiding the problem is to assume that intramolecular reactions do not occur until the moment of gel formation. A polymer of the ARB_{f-1} type, with the assumed restriction, contains one unreacted A group. The weight distribution is given by

$$W(r) = [1 - \alpha(f - 1)]r\omega_r(1 - \alpha)^{(fr-2r+1)}\alpha^{r-1} \quad (1)$$

(Flory 4, 11) where

$$\omega_r = (fr - r)!/(fr - 2r + 1)!r! \quad (2)$$

The distribution is presented for various f values in Figure 5.8.1 for $\bar{r}_n = 100$. The average degrees of polymerization are

$$\bar{r}_n = 1/[1 - \alpha(f - 1)] \quad (3)$$

$$\bar{r}_w = [1 - \alpha^2(f - 1)]/[1 - \alpha(f - 1)]^2 \quad (4)$$

$$\bar{r}_z = \{[1 - (f - 1)\alpha][1 - (f - 1)\alpha^2]$$

$$- 2[1 - (f - 1)\alpha](f - 1)\alpha^2(1 - \alpha)$$

$$+ 3[1 - (f - 1)\alpha^2](f - 1)\alpha(1 - \alpha)\}$$

$$\div [1 - (f - 1)\alpha]^2[1 - (f - 1)\alpha^2] \tag{5}$$

Inspection of these equations shows that they approach infinity as $\alpha \to \alpha_c$; hence α cannot exceed α_c under the assumption of no ring formation.

$$\alpha_c = 1/(f - 1) \tag{6}$$

At α_c the number-average molecular weight becomes infinitely large, but the area under the distribution curve must have unit area, hence the height of the curve must be infinitely thin. Therefore Figure 5.8.1 is presented only to show that a maximum in the weight fraction distribution does not exist when $f > 2$. Under these conditions:

$\bar{r}_w/\bar{r}_n$	f	$\bar{r}_n$	α	α_c
51	3	100	0.495	0.500
75.5	5	100	0.2475	0.2500

The distribution curve for $f = 3$ is shown for various conversions in Figure 5.8.2.

Now let the various B groups within a monomer have different reactivities, that is, for $f = 3$, we call the

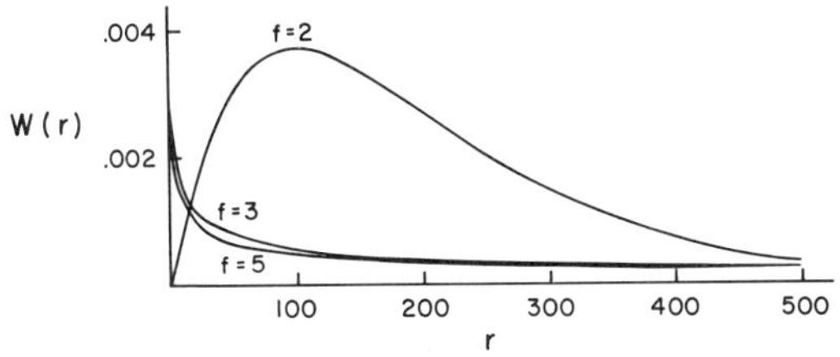

Figure 5.8.1 - The weight fraction distribution as a function of r for a homopolymer of ARB_{f-1} where A can react only with B, $\bar{r}_n = 100$. (After Flory, 11).

Curve	a	b	c
α	0.25	0.40	0.495
$\bar{r}_n$	2	5	100
$\bar{r}_w$	3.5	17	5500

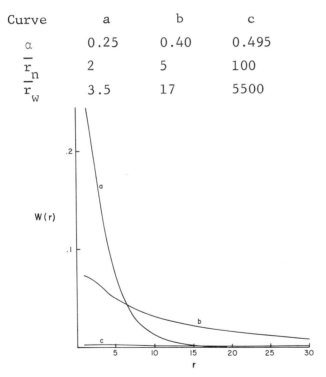

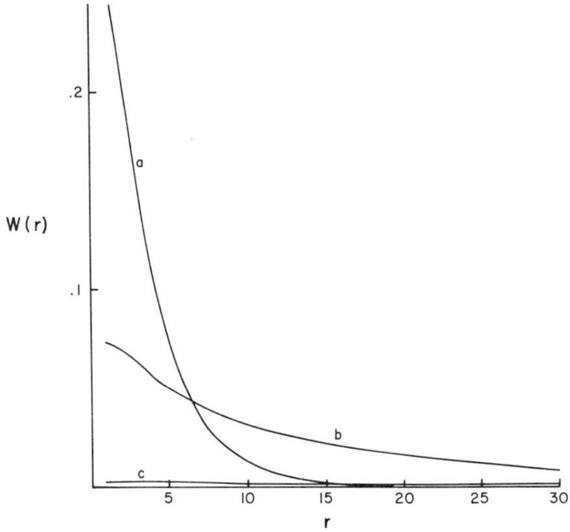

Figure 5.8.2 - The weight fraction distribution as a function of r for a homopolymer of ARB_2 and varying conversion. (Courtesy *J. Am. Chem. Soc.*)(Flory, 11).

monomer ABC, where A can react with B or C, but the last two do not react with each other. Again ring formation is prohibited, so that each molecule contains one unreacted A group. Designate i as the number of AB bonds in a molecule and j as the number of AC bonds, and the total fraction of B and C groups that have reacted as β and γ, respectively; then

$$2\alpha = \beta + \gamma$$

$$W_{ij}(r)$$

$$= (1 - \beta - \gamma)(1 - \beta)(1 - \gamma)[\beta(1 - \gamma)]^i[\gamma(1 - \beta)]^j$$

$$\times \binom{r}{i}\binom{r}{j} \qquad (r = 1 + i + j) \tag{7}$$

(Allen, 12). This distribution reduces to equation (1) when $\beta = \gamma$ and noting that

$$W(r) = \sum_{i=0}^{r-1} W_{i(r-1-i)}(r) \tag{8}$$

Thus for every molecular size r, there are $1 + i + j$ or r isomeric ways of forming the molecule. The average degrees of polymerization are given by equations (3) and (4). The function $W_{ij}(r)$ is shown in Figure 5.8.3 with $\beta = 0.2$ and $\gamma = 0.3$.

For $f \geqslant 3$, each B_i type group has a different reactivity, that is, we have a molecule of the form $AB_1B_2\ldots B_{f-1}$

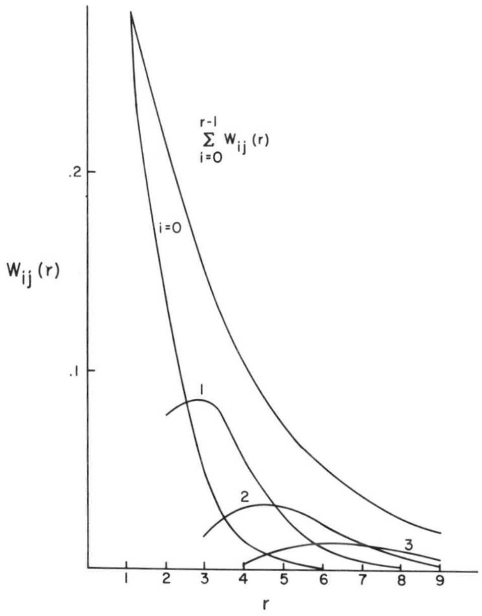

Figure 5.8.3 - The weight fraction distribution as a function of r for an ABC type molecule where A can only react with B or C. The extents of reaction are $\beta = 0.2$, $\gamma = 0.3$. Where i is the number of AB bonds and j is the number of AC bonds, $j = r - 1 - i$. The upper curve is taken from Figure 3.8.2 with $\alpha = 0.25$. (after Allen, 12).

or ABC...X. The parameters i, j, k, ... represent the number of AB_1, AB_2, AB_3, ... bonds in a particular molecule, there being f - 1 terms in the series. The various extents of reaction are related by

$$(f - 1)\alpha = \sum_{s=i}^{f-1} p_s = p_i + p_j + p_k + \cdots + p_{f-1} \qquad (9)$$

The weight distribution function is then

$$W_{ijk\ldots}(r) = \prod_{s=i}^{f-1} \binom{r}{s} p_s^{\,s}(1 - p_s)^{r-s}/\bar{r}_n \qquad (10)$$

(Erlander and French, 13) with $\bar{r}_n$ and $\bar{r}_w$ given by (3) and (4).

9. <u>Copolymer of ARB_{f-1} and AB</u>. <u>A</u> can react only with <u>B</u>. Formation of rings excluded prior to gelation.

The extension of Section 8 to the case of copolymerizing a linear polymer <u>AB</u> with branch units ARB_{f-1}, the mole fraction of branching unit being

$$\rho = ARB_{f-1}/(ARB_{f-1} + AB)$$

is given by the distribution

$$W(b,r) = [(1 - \beta)^2/\beta - (f - 2)(1 - \beta)\rho](b + r)x^b y^r z$$

(Flory, 11) where

$$x = \beta\rho(1 - \beta)^{f-2}$$

$$y = (1 - \rho)\beta$$

$$z = (r + fb - b)!/r!b!(fb - 2b + 1)!$$

$\beta = (B_o - B)/B_o$ = conversion of B units

This function is quite complex; no graphs of the distribution are given. The distribution $W(b) = \sum_{r=1}^{\infty} W(b, r)$ is similar to the distribution in Section 11.

10. Homopolymer of RA_f. Formation of rings excluded prior to gelation (4, 8, 11, 14).

The distribution function for molecules of size r is

$$W(r) = (1 - \alpha)^2 \alpha^{r-1} (1 - \alpha)^{fr-2r} fr(fr - r)!$$

$$\div (fr - 2r + 2)!r! \tag{1}$$

The average values of r are

$$\bar{r}_n = 1/(1 - \alpha f/2) \tag{2}$$

$$\bar{r}_w = (1 + \alpha)/[1 - \alpha(f - 1)] = \infty \text{ at } \alpha_c \tag{3}$$

$$\alpha_c = 1/(f - 1) \tag{4}$$

In contrast to Section 8, α can exceed α_c and $\bar{r}_n$ remains finite. The weight fraction of soluble material when α exceeds α_c is given by

$$w_f = (1 - \alpha)^2 \alpha' / (1 - \alpha')^2 \alpha \tag{5}$$

where α' is the lowest root of

$$\alpha(1 - \alpha)^{f-2} = \beta \qquad (6)$$

and β is determined by substituting the extent of conversion α into this equation. Figure 5.10.1 presents the weight-fraction distribution for RA_3 and various extents of reaction.

Curve	a	b	c
α	0.25	0.40	0.50
$\bar{r}_n$	1.6	2.5	4.0
$\bar{r}_w$	2.5	7.0	∞

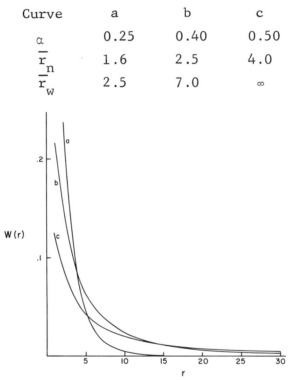

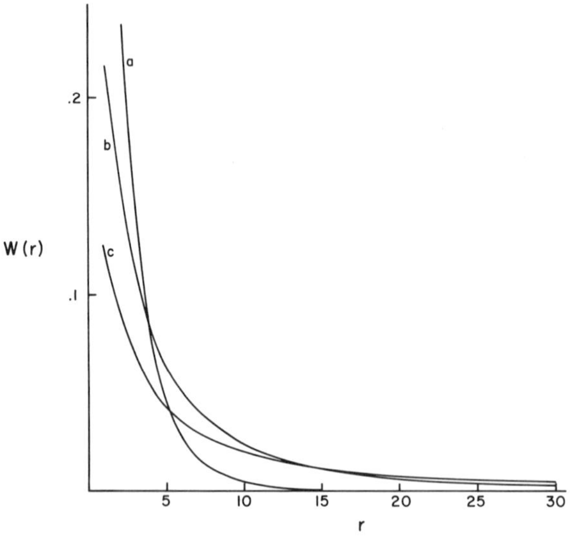

Figure 5.10.1 - The weight fraction distribution as a function of r for a homopolymer of RA_3 for various degrees of conversion. (Courtesy *Chem. Revs.*) (Flory, 16).

If RA_f is a polyepoxide, $k_i = k_p$, and no termination, the modified Poisson distribution is given by Fukui and Yamabe (15).

11\. A. Copolymer of RA_f and AA. Formation of rings excluded prior to gelation.

Let b_o = number of original RA_f groups and r_o = number of original AA groups. Then

$$\rho = fb_o/(fb_o + 2r_o) \tag{1}$$

The distribution function, in terms of r AA units and b branch units, is given by

$$W(b, r) = \{(1 - \alpha)^2/\alpha[\rho + (f/2)(1 - \rho)]\}x^b y^r z(b, r) \tag{2}$$

(Flory, 4, 8; Stockmayer, 14) where

$$x = \alpha\rho(1 - \alpha)^{f-2} \tag{3}$$

$$y = (1 - \rho)\alpha$$

$$z = (r + fb - b)!/r!b!(fb - 2b + 1)!$$

Let the parameter a be the probability that an A of one branch leads to another branch

$$a = \alpha\rho/[1 - \alpha(1 - \rho)] \tag{4}$$

Then

$$W(b) = \sum_{\substack{r=0 \\ b>0}}^{\infty} W(b, r)$$

$$= 2(1 - a)^{fb-2b+2} a^{b} \omega_b (fb - b + 1)/[a + (1 - a)\alpha]f \quad (5)$$

(Flory, 4).

$$\omega_b = f(fb - b)!/(fb - 2b + 2)!b! \quad (6)$$

$$W(0) = (1 - a/\alpha)(1 - a)/(1 - a + a/\alpha) \qquad \text{for } a < a_c \quad (7)$$

Graphs of these equations are not presented.

B. <u>Copolymer of RA_3, AA, and BB</u>. A can only react with B. Formation of rings excluded prior to gelation.

Extension of the Stockmayer distribution function and simplifying assumptions valid for long chains of randomly placed branch points (where each branch point is assumed to be completely reacted) gives

$$W_o(r) = (r/\bar{r}_n^{\,o\,2}) \exp\{-r(\gamma + 1)/\bar{r}_n^{\,o}\} \quad (8)$$

$$W_b(r) = [\gamma r^2/b(b + 2)\bar{r}_n^{\,o\,2}]W_{b-1}(r) \qquad (b \gtrsim 1,\ b/r << 1) \quad (9)$$

$$W(r) = \sum_{b=0}^{\infty} W_b(r) = 2(r/\bar{r}_n^{o\,2}) \exp\{-r(\gamma + 1)/\bar{r}_n^{o}\}$$

$$\times \sum_{b=0}^{\infty} [(\gamma r^2/\bar{r}_n^{o\,2})^b / b!(b + 2)!] \quad (10)$$

(Shultz, 17)

$$= (2/\gamma r) \exp\{-r(\gamma + 1)/\bar{r}_n^{o}\}\tilde{I}_2(2\gamma^{\frac{1}{2}} r/\bar{r}_n^{o}) \quad (11)$$

(Berger and Shultz, 18) where $\bar{r}_n^{o}$ is the $\bar{r}_n$ which would result if no branch points were present and every two RA_3 units were replaced by three AA units and maintaining the same extent of reaction:

γ = branching index = $(3/2)\bar{r}_n^{o}\,(AA)_o/(RA_3)_o$

b = number of branch points per chain

$\tilde{I}_k(x)$ = modified Bessel function of the first kind and kth order

Furthermore,

$$\bar{b}_n(r) = \sum_{b=0}^{\infty} bW_b(r) / \sum_{b=0}^{\infty} W_b(r) \quad (12)$$

$$= \gamma^{\frac{1}{2}} r \tilde{I}_3(2\gamma^{\frac{1}{2}} r/\bar{r}_n^{o})/\bar{r}_n^{o\,2}\, \tilde{I}_2(2\gamma^{\frac{1}{2}} r/\bar{r}_n^{o}) \quad (13)$$

$\bar{b}_n(r)$ is the number-average number of branch points for molecules of size r. The average degrees of polymerization are

$$\bar{r}_n = \bar{r}_n^{\,o}/(1 - \gamma/3) \tag{14}$$

$$\bar{r}_w = 2\bar{r}_n^{\,o}/(1 - \gamma) \tag{15}$$

Shultz gives a table of W(r) and $\bar{b}_n(r)$ for various values of γ. Figures 5.11.1 and 5.11.2 are constructed from the table with $\bar{r}_n = 100$ for each value of γ.

Curve	a	b	c	d	e	f
γ	0	0.2	0.4	0.6	0.8	0.96

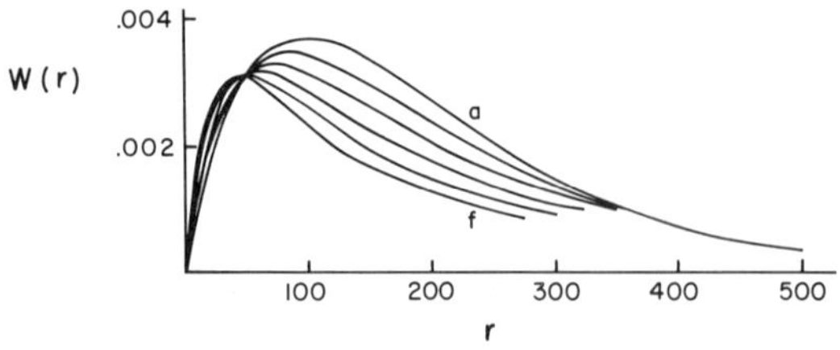

Figure 5.11.1 - Total weight fraction distribution as a function of r for various values of γ for molecules containing trifunctional units distributed randomly throughout the polymer, each trifunctional unit is assumed to be completely reacted and $\bar{r}_n = 100$. γ is the branching index. Details of the high molecular weight tail may be inaccurate because of a long interpolation. (after Shultz 17).

12. Branching Without Gelation. Copolymer of RA_f and AB. A can react only with B.

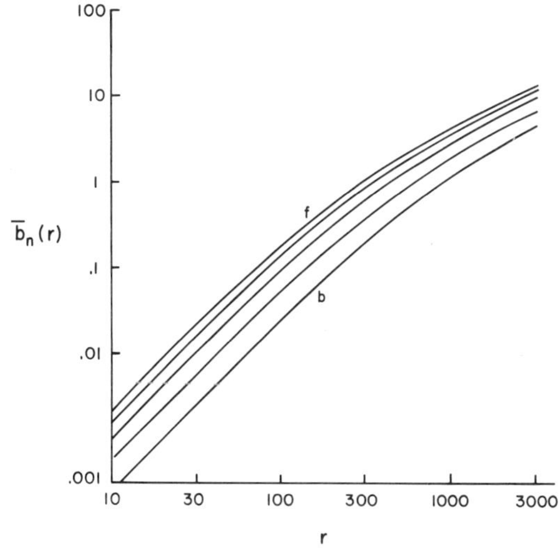

Figure 5.11.2 - Number-average number of branches per molecule as a function of r for various values of γ with $\bar{r}_n = 100$. Compare Figure 5.11.1. (After Shultz, 17).

In polymers of this type, ring formation and gelation cannot occur. The exact weight function for the entire distribution, if by-product weight can be neglected, is given by

$$W(r) = \frac{r(r + f - 2)!(1 - \alpha)^{f+1}\alpha^{r-1}}{(f - 1)!(r - 1)!(f\alpha + 1 - \alpha)} \tag{1}$$

(Schaefgen and Flory, 19) and the average degrees of polymerization by

$$\bar{r}_n = (f\alpha + 1 - \alpha)/(1 - \alpha) \tag{2}$$

$$\bar{r}_w = \frac{(f - 1)^2\alpha^2 + (3f - 2)\alpha + 1}{(f\alpha + 1 - \alpha)(1 - \alpha)} \quad (3)$$

The Schulz distribution, discussed in Chapter 1, written in the present notation is

$$W(r, f) = (r^f/f!)(1 - \alpha)^{f+1}\alpha^{r-1} \quad (4)$$

This is a reasonable approximation to equation (1) provide that we can replace $r^2 (r + f - 2)(r + f - 3)\ldots(r + 1)$ by r^f and $(f\alpha + 1 - \alpha)$ by $\underline{f}$. Figures 1.3.3 and 1.3.4 give the Schulz distribution for a variety of $\underline{f}$ values.

Note that the average degrees of polymerization are independent of the mole ratio of multifunctional unit to bifunctional unit; they depend only on the ultimate conversion achieved. When α is close to unity, we can use the following approximation for $\bar{r}_n$ and $\bar{r}_w$:

$$\bar{r}_n = f/(1 - \alpha) \quad (5)$$

$$\bar{r}_w = (1 + f)/(1 - \alpha) \quad (6)$$

$$\bar{r}_w/\bar{r}_n = 1 + 1/f \quad (7)$$

13. Gelation Conditions. Formation of rings excluded prior to gelation.

Case has presented a number of formulas for calculating the gelation condition for a whole variety of situations.

Unfortunately the equations are derived on the basis that rings may not form prior to gelation, that is, a Maxwellian demon prevents ring formation reactions until the proper extent of reaction has been achieved. Of course, this is not the case. The formulas are mainly useful to indicate the possible limit of conversion while still maintaining complete solubility. The following discussion is taken almost directly from Case (20).

A. $\underline{AA}$ reacting with B_fC_g.

Let $\underline{AA}$ react with B_fC_g, where $\underline{A}$ can react with $\underline{B}$ and $\underline{C}$, but with different velocities. $\underline{B}$ cannot react with $\underline{C}$. Rings are not formed prior to gelation.

The gelation condition is then

$$1 = [\alpha/(f\beta + g\gamma)][f(f - 1)\beta^2 + 2fg\beta\gamma + g(g - 1)\gamma^2] \quad (1)$$

An example would be adipic acid reacting with trimethylolpropane or with glycerol (for which all hydroxyl groups are considered to be of the same reactivity). Here $\underline{AA}$ reacts with RB_3 and

$$1 = \frac{\alpha}{3\beta}\ [(3)(2)\beta^2] = 2\alpha\beta \qquad \text{for gelation} \quad (2)$$

Another example is adipic acid reacting with pentaerythritol for which all hydroxyl groups are considered to have the same reactivity. Here $\underline{AA}$ reacts with RB_4 and

$$1 - \frac{\alpha}{4\beta}\ [(4)(3)\beta^2] = 3\alpha\beta \qquad \text{for gelation} \quad (3)$$

Still another example is adipic acid reacting with glycerol for which the secondary hydroxyl group is considered to be of different reactivity from that of the primary groups. Here <u>AA</u> reacts with RB_2C and

$$1 = \left(\frac{\alpha}{2\beta + \gamma}\right)[2\beta^2 + 4\beta\gamma] \qquad \text{for gelation} \tag{4}$$

These formulas specify the condition for gelation, when the weight-average molecular weight goes to infinity. Because <u>B</u> and <u>C</u> can react with different velocities, the fractions of α, β, and γ will depend on the kinetics of the system. If the rate constants either are known or can be assumed, with the aid of the mass balance conditions, values of α, β, and γ at gelation can be found in a straightforward way. Thus let there be equivalent amounts of acid and glycerol, so that

$$3\alpha = 2\beta + \gamma$$

The reactions of <u>A</u> with <u>B</u> and <u>C</u> yield

$$-dB/dt = k_1AB$$

$$-dC/dt = k_2AC \tag{5}$$

so that

$$dB/k_1B = dC/k_2C$$

Thus

$$(B/B_o) = (C/C_o)^{k_1/k_2} = (1 - \beta) = (1 - \gamma)^{k_1/k_2} \qquad (6)$$

and the condition for gelation $1 = 2\alpha\beta(\beta + 2\gamma)/(2\beta + \gamma)$. These equations can now be solved for α, β, and γ when k_1/k_2 is assumed:

k_1/k_2	α	β	γ
1	0.707	0.707	0.707
2	0.711	0.794	0.546
5	0.729	0.905	0.376
10	0.742	0.968	0.291

B. AA, BB, and C reacting with DDE and FF.

The groups A, B, and C individually can react with D, E, and F, but with different velocities. Let

$$\nu = (BB)_o/(AA)_o$$

$$\rho = (C)_o/2(AA)_o \qquad (7)$$

$$\pi = 2(FF)_o/(DDE)_o$$

The symbols α, β, γ, δ, ϵ, ζ, refer to the extent of reaction of the groups A, B, C, D, E, and F. The gelation condition is

$$1 = \left[\frac{\alpha^2 + \nu\beta^2}{\alpha + \nu\beta + \rho\gamma}\right]\left[\frac{4\epsilon\delta + 2\delta^2 + \pi\zeta^2}{\epsilon + 2\delta + \pi\zeta}\right] \qquad (8)$$

An example is adipic acid reacting with glycerol, in which $\underline{AA}$ reacts with $\underline{DDE}$. Here:

$$1 = \frac{\alpha^2(4\epsilon\delta + 2\delta^2)}{\alpha(\epsilon + 2\delta)} = \frac{\alpha}{(2\delta + \epsilon)}(2\delta^2 + 4\delta\epsilon) \tag{9}$$

which is the same as that given earlier.

Another example is the reaction of adipic acid with glycerol and glycol, in which the groups in glycol are assumed to be of the same reactivity as the primary groups in glycerol. Here

$$\underline{AA} \text{ reacts with } \underline{DDE} + \underline{DD} \tag{10}$$

$\pi/2$ = the ratio of glycol to glycerol

$$1 = \frac{\alpha^2(4\epsilon\delta + 2\delta^2 + \pi\delta^2)}{\alpha(\epsilon + 2\delta + \pi\delta)} = \frac{\alpha[4\epsilon\delta + (2 + \pi)\delta^2]}{\epsilon + (2 + \pi)\delta} \tag{11}$$

Still another example is the same as those above, except that the primary groups in the glycerol are assumed to have different reactivities from those in glycol. Here $\underline{AA}$ reacts with $\underline{DDE}$ and $\underline{FF}$

$\pi/2$ = the ratio of glycol to glycerol

$$1 = \frac{\alpha^2(4\epsilon\delta + 2\delta^2 + \pi\zeta^2)}{\alpha(\epsilon + 2\delta + \pi\zeta)} = \frac{\alpha(4\epsilon\delta + 2\delta^2 + \pi\zeta^2)}{\epsilon + 2\delta + \pi\zeta} \tag{12}$$

Other simplifications may be made.

C. AAB, RC_4, and GG reacting with DE and F.

Let AAB and RC_4 and GG react with DE and F, where DE is an anhydride or similar material. Group D must react first and may have a different velocity of reaction from group E. A, B, C, and G may react with different velocities.

$$\nu = 4(\mathrm{CCCC})_o/(\mathrm{AAB})_o$$

$$\rho = (\mathrm{F})_o/(\mathrm{DE})_o \tag{13}$$

$$\pi = 2(\mathrm{GG})_o/(\mathrm{AAB})_o$$

The symbols δ' and ε' are primed to show that they refer to an anhydride and that δ' must be greater than or equal to ε'. The gelation condition is

$$1 = \left[\frac{2\varepsilon'}{\delta' + \varepsilon' + \rho\zeta}\right]\left[\frac{4\alpha\beta + 2\alpha^2 + 3\gamma^2\nu + \pi\eta^2}{2\alpha + \beta + \nu\gamma + \pi\eta}\right] \tag{14}$$

D. AB and CD reacting with EEF and GG.

Let AB and CD react with EEF and GG where AB and CD are anhydrides or similar materials and A must react before B and C must react before D. A, B, C, and D react with different velocities.

$$\nu = (\mathrm{CD})_o/(\mathrm{AB})_o$$

$$\rho = 2(\mathrm{GG})_o/(\mathrm{EEF})_o \tag{15}$$

The α', β', γ', and δ' are primed to show that they are derived from anhydrides. The gelation condition is

$$1 = \left[\frac{2\nu\delta' + 2\beta'}{\nu(\gamma' + \delta') + \alpha' + \beta'}\right]\left[\frac{2\epsilon^2 + 4\epsilon\zeta + \rho\eta^2}{2\epsilon + \zeta + \rho\eta}\right] \tag{16}$$

E. AA and BC reacting with DDE and FG.

Let AA and BC react with DDE and FG where BC is an anhydride. B reacts first, and FG is similar to an unsymmetrical glycol. Different species may react with different velocities.

$$\nu = (BC)_o/2(AA)_o$$

$$\rho = (FG)_o/(DDE)_o \tag{17}$$

The gelation condition is

$$1 = \left[\frac{2\alpha^2 + 2\nu\gamma'}{2\alpha + \nu(\beta' + \gamma')}\right]\left[\frac{2\delta^2 + 4\delta\epsilon + 2\rho\zeta\eta}{2\delta + \epsilon + \rho(\zeta + \eta)}\right] \tag{18}$$

14. Ring Formation in Linear Polymers

In the preceding sections, ring formation has been assumed not to occur prior to gelation. However, ring formation can occur even in the polymerization of linear, bifunctional monomers, because the "front end" of the molecule can react with the "back end" to produce polymer molecules without end groups. Let the weight fraction of rings of size $\underline{r}$ be called W(r)(rings); then for monomers of type AA

$$W(r)(\text{rings}) = (BM/c)(\alpha')^r/r^{3/2} \tag{1}$$

(Jacobson and Stockmayer, 21) where $B = (3/2\pi\nu)^{3/2}/2b^3N_o$. The $\underline{M}$ is the molecular weight of the repeat unit, $\underline{c}$ is the concentration in grams per milliliter, ν is the number of atoms in the repeat unit backbone, $\underline{b}$ is the effective link length of each atom, corrected for bond angles and hindrance to free rotation, N_o is Avogadro's number, and α' is a revised extent of reaction defined by

$$1 - \alpha' = (1 - \alpha)/(1 - w_r) \tag{2}$$

where w_r is the weight fraction of rings in the system

$$w_r = \Sigma W(r)(\text{rings}) \tag{3}$$

and α is the overall extent of conversion. To find w_r, first define the sum

$$\varphi(\alpha', s) = \sum_{n=1}^{\infty} (\alpha')^n n^{-s} \tag{4}$$

which is given in Table 5.14.1. The average values of $\underline{r}$ and w_r are given by

$$\bar{r}_n \text{ (chain fraction)} = 1/(1 - \alpha') \tag{5}$$

$$\bar{r}_n \text{ (ring fraction)} = \varphi(\alpha', 3/2)/\varphi(\alpha', 5/2) \tag{6}$$

$$w_r = (BM/c)\varphi(\alpha', 3/2) \tag{7}$$

TABLE 5.14.1

Values of Ring Functions

$$\varphi(\alpha', s) = \sum_{n=1}^{\infty} (\alpha')^n n^{-s}$$

α'	$\varphi(\alpha' 5/2)$	$\varphi(\alpha' 3/2)$	$\varphi(\alpha' 1/2)$
0.1	0.102	0.104	0.108
0.2	0.208	0.216	0.234
0.3	0.318	0.338	0.385
0.4	0.434	0.473	0.570
0.5	0.555	0.625	0.806
0.6	0.684	0.798	1.122
0.7	0.822	1.003	1.580
0.75	0.895	1.122	1.903
0.8	0.972	1.258	2.338
0.85	1.053	1.418	2.970
0.9	1.139	1.614	4.022
0.95	1.232	1.884	6.377
0.97	1.274	2.038	8.702
0.99	1.317	2.272	16.22
0.995	1.329	2.366	23.30
0.997	1.334	2.422	30.88
0.999	1.339	2.501	54.58
1	1.341	2.612	∞

(Courtesy of J. Chem. Phys.)(Jacobson and Stockmayer, 21).

$$\bar{r}_w \text{ (ring fraction)} = \varphi(\alpha', 1/2)/\varphi(\alpha', 3/2) \tag{8}$$

$$\bar{r}_w \text{ (chain fraction)} = (1 + \alpha')/(1 - \alpha') \quad (9)$$

$$\bar{r}_w \text{ (total)} = w_r \bar{r}_w \text{ (rings)} + (1 - w_r)\bar{r}_w \text{ (chains)} \quad (10)$$

The distribution of rings for the monomers of type AB is found by replacing the parameter B by 2B in equations (1) and (7).

Ring formation in the system AA and BB is somewhat more difficult to formulate. Because rings can only be formed from an AABB unit, let this be the repeat unit and r is the number of such units; then

$$W(r)\text{(rings)} = (2BM/c)(\alpha')^{2r}/(r)^{3/2} \quad (11)$$

(Jacobson and Stockmayer, 21) where the terms are defined earlier but are adjusted to the new repeat unit size.

When the moles of AA exactly equal the moles of BB in the system,

$$\bar{r}_n \text{ (chain fraction)} = 1/(1 - \alpha') \quad (12)$$

$$\bar{r}_n \text{ (ring fraction)} = 2\varphi(\alpha'^2, 3/2)/\varphi(\alpha'^2, 5/2) \quad (13)$$

$$w_r = (2BM/c)\varphi(\alpha'^2, 3/2) \quad (14)$$

$$\alpha = \alpha' + (1 - \alpha')w_r \quad (15)$$

$$\bar{r}_w \text{ (chains)} = (1 + \alpha')/(1 - \alpha') \quad (16)$$

$$\bar{r}_w \text{ (rings)} = 2\varphi(\alpha'^2, 1/2)/\varphi(\alpha'^2, 3/2) \tag{17}$$

$$\bar{r}_w \text{ (total)} = w_r\bar{r}_w \text{ (rings)} + (1 - w_r)\bar{r}_w \text{ (chains)} \tag{18}$$

When <u>AA</u> is in excess and <u>BB</u> has reacted completely so that each linear molecule has <u>A</u> units at both ends, the revised extent of reaction α'' is

$$1 - \alpha'' = 2(1 - \alpha)/[(1 - \alpha) + (1 - w_r)] \tag{19}$$

$$\alpha = 2\rho/(1 + \rho) \text{ and } \rho \text{ is the ratio of } \underline{BB} \text{ to } \underline{AA} \text{ and } \rho < 1 \tag{20}$$

$$\bar{r}_n \text{ (chains)} = (1 + \alpha'')/(1 - \alpha'') \tag{21}$$

$$\bar{r}_n \text{ (rings)} = 2\varphi(\alpha'', 3/2)/\varphi(\alpha'', 5/2) \tag{22}$$

$$w_r = (2BM/c)\varphi(\alpha'', 3/2) \tag{23}$$

$$\bar{r}_w \text{ (chains)} = [1 + 6\alpha'' + (\alpha'')^2]/[1 - (\alpha'')^2] \tag{24}$$

$$\bar{r}_w \text{ (rings)} = 2\varphi(\alpha'', 1/2)/\varphi(\alpha'', 3/2) \tag{25}$$

The total weight fraction of rings, w_r, equation (14) is plotted against the total conversion α obtained from equation (15) in Figure 5.14.1. From the lower curve to the upper curve, the concentration of the polymer <u>decreases</u>, that is, more rings are formed at high dilution. Furthermore, the graph shows that the total concentration of rings in a high molecular weight condensation polymer prepared in the melt (no solvent) will be practically independent of the overall conversion. That is, the cyclic oligomer content should be practically independent of the molecular weight of the polymers.

The distribution for the chain fraction, given in earlier chapters is not rigorously correct; the equation should be modified to

$$W(r)\ (\text{chains}) = (1 - w_r)(1 - \alpha')^2 r(\alpha')^{r-1} \qquad (26)$$

(Flory, 4). Consider an example where 99% of the active groups have reacted and 2 weight % of the polymer is in the form of rings. From equation (15), $\alpha' = 0.98980$. The curve for equation (26) is practically indistinguishable from the Schulz-Flory distribution (compare Figure 5.14.2). The distribution of rings, is based on the statistical probability that the "front end" can reach around to the "middle" of a linear polymer to form a ring. In some systems, it is sterically difficult to form small rings; in this case there should be a maximum in the distribution of rings with increasing size.

The problem of equilibrium constant for the reaction

$$\text{chains of size } r \rightleftharpoons \text{chains of size } (r - g) + \text{rings of size } g \qquad (27)$$

have been considered by Jacobson and Stockmayer (21),

Curve	a	b	c
2BM/c	0.01	0.1	1.0

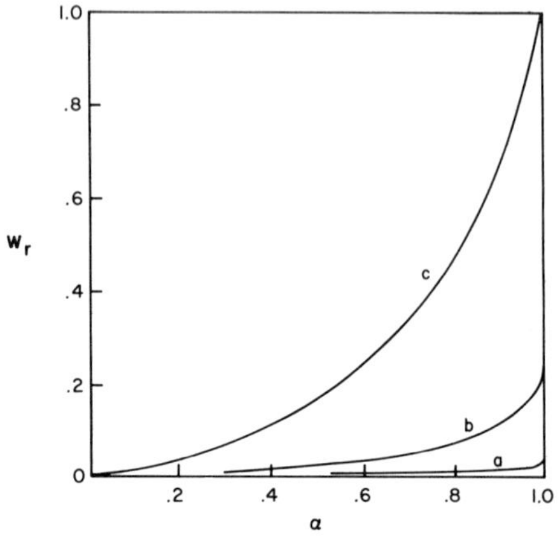

Figure 5.14.1 - Total weight fraction of rings, w_r (equation 14) as a function of conversion α and reciprocal concentration, 2BM/c (Courtesy J. Chem. Phys.) (Jacobson and Stockmayer, 21).

r	100	200	300	400	500
$W(r) \times 10^5$ (chains)	370	265	143	68	31
$W(r) \times 10^5$ (Schulz-Flory)	370	271	149	73	33

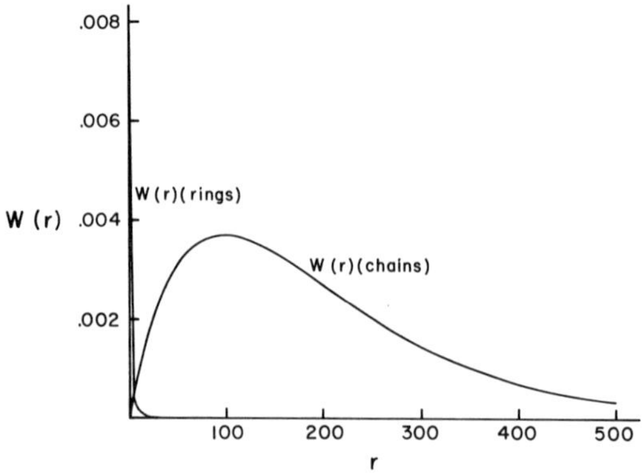

Figure 5.14.2 - The weight fraction distribution as a

function of r for an equimolar mixture of AA and BB type monomers $\bar{r}_n = 100$, $w_r = 0.02$. W(r)(chains) is practically indistinguishable from the Schulz-Flory distribution (after Flory, 4).

Flory and Semlyen (22), and Levy and Van Wazer (23). In general the equilibrium constant is proportional to $g^{-5/2}$.

15. Gelation of RA_f When Rings are Permitted Prior to Gelation

Harris (24) has considered the case of polymers of type RA_f where ring formation is permitted to form prior to gelation. He finds that gelation is predicted for all concentrations at all nonzero extents of reaction if he neglects geometric restrictions on ring formation. Therefore he assumes that the number of ends available for ring formation with a given end depends only on the number of ends per unit weight of polymer. In other words, each end of the noncyclic chain is assumed to have a constant number, σ, of ends within range for possible closure. The number σ does not increase with the size of the polymer. We define the "effective extent of reaction," α, by the relations

$$\bar{r}_n = 1/(1 - \alpha f/2) \tag{1}$$

$$\bar{r}_w = (1 + \alpha)/[1 - \alpha(f - 1)] \tag{2}$$

and the "actual extent of reaction," p, is defined by $1 - p = (A)/(A)_o$ or $p = 2J/fN$ where J is the number of bonds in the polymer which may contain rings and N is the original number of monomers. The condition for gelation is given by

$$\alpha_c = 1/(f - 1) \tag{3}$$

The volume per monomer, a reciprocal initial concentration, is given by

$$\frac{AV}{N} = \frac{(1 - \alpha)^2 fu(1 + u)^{\frac{1}{2}}}{2^{\frac{1}{2}}\alpha e^{1-u}} \tag{4}$$

where A is a constant involving the root mean square end-to-end distance, h_o of a chain of r units,

$$A = (\sigma/8re\langle h_o{}^3\rangle)(3/\pi)^{3/2} \tag{5}$$

and

$$u = (p - \alpha)/(1 - \alpha) \tag{6}$$

A plot of p versus AV/N is given in Figure 5.15.1. At sufficiently high volumes, the actual extent of conversion can approach unity without any gel being formed, because internal ring formation predominates. The critical volume for gelation is given by

$$\frac{AV_c}{N} = \frac{(f - 2)^2 f}{f - 1} \tag{7}$$

Curve	a	b	c
α	0.25	0.40	0.50
$\bar{r}_n$	1.6	2.5	4.0

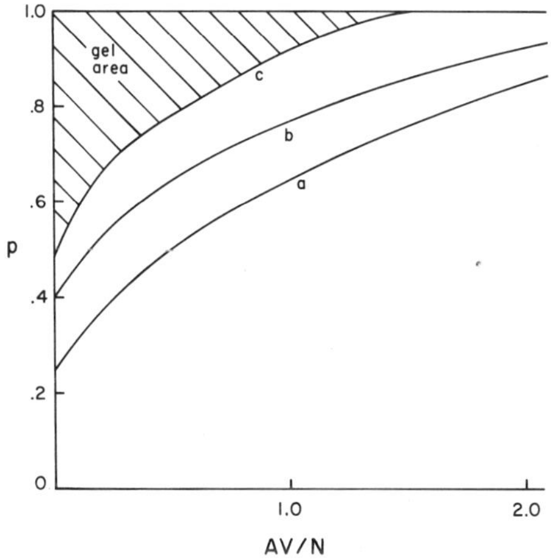

Figure 5.15.1 - The "actual" extent of reaction, p, as a function of AV/N (a reciprocal concentration) and the "effective" extent or reaction, α. The area to the right of the cross-hatched region is that of complete solubility. (Courtesy J. Chem. Phys.) (Harris, 24).

On the basis of graph theory, Bruneau (25) defines the following parameters. Let P be the total number of polymer molecules and

$$\nu = J + P - N \tag{8}$$

where ν is the cyclomatic number of the system. Furthermore

$$\nu = (p \frac{\bar{f}}{2} - 1)N + P \tag{9}$$

where $\overline{f}$ is the average functionality of the system. The average number of rings per polymer molecule is given by

$$\nu/P = 1 - \overline{r}_n[1 - (p\overline{f}/2)] \tag{10}$$

$$= 1 - [1 - (p\overline{f}/2)]/[1 - (\alpha\overline{f}/2)]$$

by substituting equation (1). The number of rings per molecule as a function of extent of conversion is shown in Figure 5.15.2.

Curve	a	b	c
α	0.25	0.40	0.50
$\overline{r}_n$	1.6	2.5	4.0

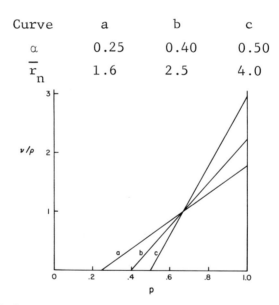

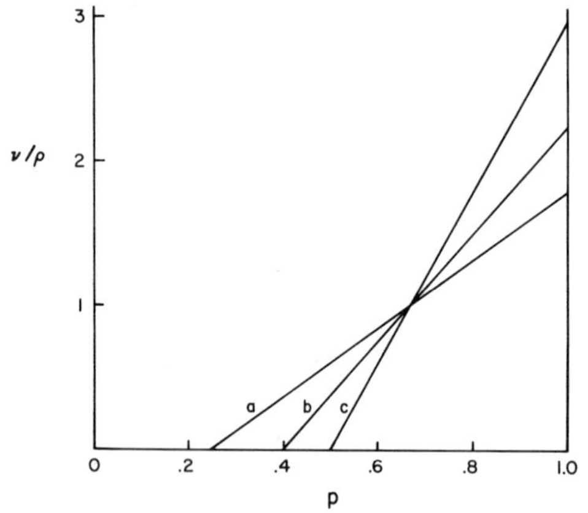

Figure 5.15.2 - The number of rings per molecule as a function of "actual" extent of reaction, p, and the "effective" extent of reaction, α, for trifunctional monomers. (after Bruneau, 25).

16. <u>Homopolymer of RA_f</u>. Rings permitted prior to gelation.

In Sections 6 and 7 we considered the frequency distribution for molecules in the form $m\{N_s\}$ or $m\{M_i, N_j\}$; the braces denote a series of N_s existing for a particular molecule $s = 1$ to ∞ and $\underline{N}$ is the number or amount of the entity $\underline{s}$ appearing in the molecule in question. Hoeve (26) defines M_s as the number of rings which contain $\underline{s}$ units. Thus the set for a polymer molecule with two rings is shown schematically, with $f = 3$; a curved line indicates intramolecular reaction.

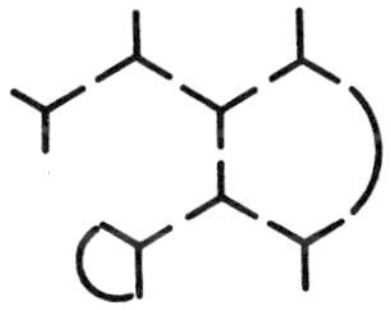

The set $\{M_s\}$ for this molecule is $\{2_0, 1_1, 0_2, 0_3, 1_4, 0_5, \ldots\}$. We read this to mean that two monomers are not involved in rings, there is one ring with a unit ring size, there are no rings containing either two or three monomers, and one ring contains four monomers. Again, no distinction is made as to the assorted possible stereoisomers; there are nine distinguishable isomers of the molecule shown. All of these stereoisomers are considered as structural variations of the given molecule.

The parameter M_o is the number of monomers not involved in rings in a particular molecule. The various M_o's are distributed among unreacted monomer, linear chains, branched chains that do not contain rings, and segments within molecules that do contain rings (i.e., the linear section between two rings).

The ΣsM_s is the number of monomers involved in rings in a given molecule; $m\{M_s\}$ is the number of molecules which make up the set $\{M_s\}$, that is for our schematic molecule, it is the number of heptamers which have the structure $\{2_0, 1_1, 1_4\}$.

The total number of monomers present is given by

$$N = \Sigma(M_o + \Sigma sM_s)m\{M_s\} \tag{1}$$

$$= N_o + \Sigma sN_s \tag{2}$$

where the various M_i in the parenthesis of equation (1) are the same as those appearing in the braces. The N_o is the total number of monomers not in rings and ΣsN_s is the total number of monomers in rings.

The total number of reacted groups is given by pfN, while the number of groups involved in ring formation is given by $2\Sigma sN_s$. The total number of groups capable of reaction and derived from the total nonring monomers is fN_o and the total number of groups capable of reaction from monomers, which are involved in rings is $(f - 2)\Sigma sN_s$. With these definitions we can define α as the extent of reaction of available groups which are not involved in rings:

$$\alpha = (pfN - 2\Sigma sN_s)/[(f - 2)\Sigma sN_s + fN_o] \tag{3}$$

The desired distribution is

$$m\{M_s\} = \frac{A\{(f - 1)M_o + (f - 2)\Sigma sM_s - \Sigma M_s\}!(fx)^{M_o}}{\{(f - 2)(M_o + \Sigma sM_s) - 2\Sigma M_s + 2\}!M_o!}$$

$$\times \prod_{s=1}^{\infty} \frac{\{(f - 2)sY_s\}^{M_s}}{M_s!} \tag{4}$$

(Hoeve, 26) where

$$Y_s = \frac{kp}{2(1 - p)^2} \left[\frac{f!x}{(f - 2)!}\right]^s / s^{5/2} \tag{5}$$

The $\underline{k}$ is a constant of proportionality which can be either calculated by the method of Haward (27) or determined from experiment as described below:

$$x = [\alpha(1 - \alpha)^{f-2}/f]B(f, \alpha, p) \tag{6}$$

$$B(f, \alpha, p) = \frac{1 - f(p - \alpha)/2(1 - \alpha)}{1 - (p - \alpha)/(1 - \alpha)} \tag{7}$$

$$A = [fN_o + (f - 2)\Sigma sN_s](1 - \alpha)^2/\alpha \tag{8}$$

To obtain experimental measures of $\underline{k}$ and α, recall the definition of Section 14, which is detailed in Table 5.14.1:

$$\varphi(a, s) = \sum_{n=1}^{\infty} a^n n^{-s} \tag{9}$$

Then determine the extent of reaction, $\underline{p}$, at the gel point and solve the following sets of equations for $\underline{k}$ and α:

$$B(f, \alpha, p) + \frac{kp(f - 2)(1 - \alpha)^2}{2\alpha(1 - p)^2} \varphi(a, 3/2) = 1 \qquad (10)$$

$$fB(f, \alpha, p) + \frac{kp(f - 2)^2(1 - \alpha)^2}{2\alpha(1 - p)^2} \varphi(a, 1/2) = 1 + 1/\alpha \qquad (11)$$

where

$$a = (f - 1)\alpha B(f, \alpha, p) \qquad (12)$$

The number-average degree of polymerization is

$$\bar{r}_n = 1/\left[1 - \frac{f(p - \alpha)}{2(1 - \alpha)} - \frac{\alpha f(1 - p)}{2(1 - \alpha)} + \frac{kpf(1 - \alpha)}{2\alpha(1 - p)} \varphi(a, 5/2)\right] \qquad (13)$$

17. Gelation Condition for RA_f, AA, and BB When Rings Are Allowed to Form Prior to Gelation

$$\rho = fA_f/(fA_f + 2AA) \qquad (1)$$

The gelation condition is

$$1 = \rho(f - 1)\alpha\beta[1 - \theta\varphi(1, s)][1/(1 - z)]$$
$$-a\theta[1/(1 - z) - 1]$$
$$-b\theta(1 - a\theta)[1/(1 - z) - 1 - z]$$

$$-c\theta(1 - a\theta)(1 - b\theta)[1/(1 - z) - 1 - z - z^2]$$

$$-d\theta(1 - a\theta)(1 - b\theta)(1 - c\theta)[1/(1 - z) - 1 - z - z^2 - z^3]$$

$$-\ldots \qquad (2)$$

(Kilb, 28) where

$$z = (1 - \rho)\alpha\beta$$

$$a = \varphi(1, s) - 1$$

$$b = a - 2^{-s}$$

$$c = b - 3^{-s}$$

$$d = c - 4^{-s} \qquad (3)$$

$s = 5/2$ if ring formation is reversible

$= 3/2$ if ring formation is not reversible.

$$\theta = BM/Ac$$

$$A = N(1 - w_r)(1 - \alpha')^2/\alpha'$$

N = number of repeat units present

Other terms are defined in set 14.

18. References

1. C. H. Bamford and H. Tompa, "The Calculation of Molecular Weight Distributions from Kinetic Schemes," Trans. Faraday Soc., 50, 1097 (1954).
2. W. W. Graessley, H. Mittelhauser, and R. Maramba, "Molecular Weight Distribution in Free Radical Polymers. The Effect of Branching on Distribution Breadth, and the Prediction of Gel Points in Diene Polymerizations," Makromol. Chem., 86, 129 (1965).

2b. W. W. Graessley, "Molecular Weight Distribution in Branched Polymers--The Influence of Reactor System in the Distribution Breadth," A.I.Ch.E-I. Chem. E. Joint Meeting, London, June, 1965, 3 16.

3. J. K. Beasley, "The Molecular Structure of Polyethylene. IV. Kinetic Calculation of the Effect of Branching on Molecular Weight Distribution," J. Am. Chem. Soc., 75, 6123 (1953).
4. P. J. Flory, Principles of Polymer Chemistry, Cornell University Press, Ithaca, N. Y., 1953.
5. W. H. Stockmayer, "Molecular Distribution in Condensation Polymers," J. Polymer Sci., 9, 69 (1952); 11, 424 (1953).
6. I. Nakamura, R. Yokouchi, T. Ito, D. Miura, and K. Fujii, "Molecular Distribution of Condensation Copolymers--Copolymers Including Both Branching Agent and Chain-Terminator," Kobunshi Kagaku, 21, 553 (1964)
7. W. H. Stockmayer, "Theory of Molecular Size Distribution and Gel Formation in Branched Polymers. II.

General Cross Linking," J. Chem. Phys., 12, 125 (1944).

8. P. J. Flory, "Molecular Size Distribution in Three Dimensional Polymers. II. Trifunctional Branching Units," J. Am. Chem. Soc., 63, 3091 (1941).
9. P. J. Flory, "Molecular Size Distribution in Three Dimensional Polymers. III. Tetrafunctional Branching Units," J. Am. Chem. Soc., 63, 3096 (1941).
10. A. Amemiya, "Change of Molecular Size Distribution of Polymers Induced by Cross-linking. I. Uniform Distribution. II. Poisson Distribution," J. Phys. Soc., Japan, 23, 1394, 1402 (1967).
11. P. J. Flory, "Molecular Size Distribution in Three Dimensional Polymers. VI. Branched Polymers Containing ARB_{f-1} Type Units," J. Am. Chem. Soc., 74, 2718 (1952).
12. E. S. Allen, "A Probability Theory for the Condensation of ARB_2 Units," J. Polymer Sci., 21, 349 (1956).
13. S. Erlander and D. French, "A Statistical Model for Amylopectin and Glycogen. The Condensation of ARB_{f-1} Units," J. Polymer Sci., 20, 7 (1956).
14. W. H. Stockmayer, "Theory of Molecular Size Distribution and Gel Formation in Branched-Chain Polymers," J. Chem. Phys., 11, 45 (1943).
15. K. Fukui and T. Yamabe, "Statistical Theory of the Polymerization of Polyepoxide Monomers," J. Polymer Sci., A2, 3743 (1964).
16. P. J. Flory, "Fundamental Principles of Condensation Polymerization," Chem. Rev., 39, 137 (1946).

17. A. R. Shultz, "Polydisperse Polymers with Random Trifunctional Branching. I. Evaluation of the Extent of Branching," J. Polymer Sci., A3, 4211 (1965).
18. H. L. Berger and A. R. Shultz, "Polydisperse Polymers with Random Trifunctional Branching. II. Weight-Averaging of Chosen Molecular Size-Dependent Function and a Restatement of the Approximate Distribution Function," J. Polymer Sci., A3, 4227 (1965).
19. J. R. Schaefgen and P. J. Flory, "Synthesis of Multichain Polymers and Investigation of Their Viscosities J. Am. Chem. Soc., 70, 2709 (1948).
20. L. C. Case, "Molecular Distributions in Polycondensations Involving Unlike Reactants. I. Gelation," J. Polymer Sci., 26, 333 (1957).
21. H. Jacobson and W. H. Stockmayer, "Intramolecular Reaction in Polycondensations. I. The Theory of Linear Systems," J. Chem. Phys., 18, 1600 (1950).
22. P. J. Flory and J. A. Semlyen, "Macrocyclization Equilibrium Constants and the Statistical Configuration of Poly(dimethylsiloxane) Chains," J. Am. Chem Soc., 88, 3209 (1966).
23. R. M. Levy and J. R. Van Wazer, "Molecular Distributions at Equilibrium. V. Statistical-Mechanical Treatment of Ring-Chain Equilibria," J. Chem. Phys., 45, 1824 (1966).
24. F. E. Harris, "Ring Formation and Molecular Weight Distributions in Branched-Chain Polymers. I," J. Chem. Phys., 23, 1518 (1955).

25. C. M. Bruneau, "Theory of Stochastic Graphs Applied to the Synthesis and Random Degradation of Multifunctional Macromolecular Compounds," Thesis, Paris, 1966.

26. C. A. J. Hoeve, "Molecular Weight Distribution of Thermally Polymerized Triglyceride Oils. II. Effect of Intramolecular Reaction," J. Polymer Sci., 21, 11 (1956).

27. R. N. Haward, "Polymerization of Diallyl Phthalate," J. Polymer Sci., 14, 535 (1954).

28. R. W. Kilb, "Dilute Gelling Systems. I. The Effect of Ring Formation on Gelation," J. Phys. Chem., 62, 969 (1958).

Chapter 6

Equilibrium Polymerization

Contents

1. Introduction and the Thermodynamic Distributions

Most of the distributions so far have been derived without consideration of reversible reactions and regardless of whether depolymerization occurs simultaneously with polymerization. In this chapter we consider systems in equilibrium, such as

$$rM \rightleftharpoons P_r \tag{1}$$

The change in the Gibbs free energy for polymerization is

$$\Delta F_p = \Delta H_p - T\ \Delta S_p \tag{2}$$

and is zero for a system in equilibrium; hence we have the critical temperature

$$T_c = \Delta H_p/\Delta S_p = \Delta H_p/(\Delta S_p{}^o + R\ \ln a_M) \tag{3}$$

where $\Delta S_p{}^o$ is the standard entropy of polymerization for a system when the monomer activity is unity. If both ΔH_p and ΔS_p are negative, T_c is a "ceiling temperature" above which P_r is thermodynamically unstable. Likewise, if ΔH_p and ΔS_p are both positive, T_c is a "floor temperature" below which it is thermodynamically impossible to convert rM into P_r. As the monomer activity, a_M, can change with environment, so also can the ceiling or floor temperatures be altered with a change in solvent or monomer concentration (O'Driscoll, 1).

The molecular weight distribution of P_r will depend upon the thermodynamic definition of the final product. If the change in Gibbs free energy of polymerization is a linear function of molecular size <u>r</u>,

$$\Delta F_p = \eta + \varsigma\ r \tag{4}$$

(η and ς dependent on pressure and temperature), then two cases can be considered.

A. The monomer is polymerized to a perfectly ordered state (a solid). This gives the Schulz-Flory "most probable" distribution

$$F(r) = (1 - p)p^{r-1} \tag{5}$$

$$W(r) = (1 - p)^2 p^{r-1} \tag{6}$$

B. The monomer is polymerized to a randomly ordered state (a liquid); then

$$F(r) = (1/\ln \bar{r}_w)(1/r)(1 - 1/\bar{r}_w)^r \tag{7}$$

(Lundberg, 2)

$$\bar{r}_n = (\bar{r}_w - 1)/\ln \bar{r}_w \tag{8}$$

$$\bar{r}_w = \exp(1 + \eta/RT) + 1 \tag{9}$$

With the substitution

$$p = 1 - 1/\bar{r}_w \tag{10}$$

we find

$$W(r) = (1 - p)p^{r-1} \tag{11}$$

which is the right-hand side of equation (5). Thus

equation (7) can also be written as

$$F(r) = \bar{r}_n (1 - p) p^{r-1} / r \tag{12}$$

F(r) and W(r) for the randomly ordered state are presented in Figure 6.1.1. This is a broader distribution than that of equations (5) and (6). The parameters $\bar{r}_n$ and $\bar{r}_w/\bar{r}_n$ are plotted as a function of $\bar{r}_w$ in Figure 6.1.2.

The Lundberg distribution is also a result of extreme deviations from ideality. Thus the Schulz-Flory distribution can be derived not only from ideal solution theory at infinite dilution, but also from the Flory-Huggins

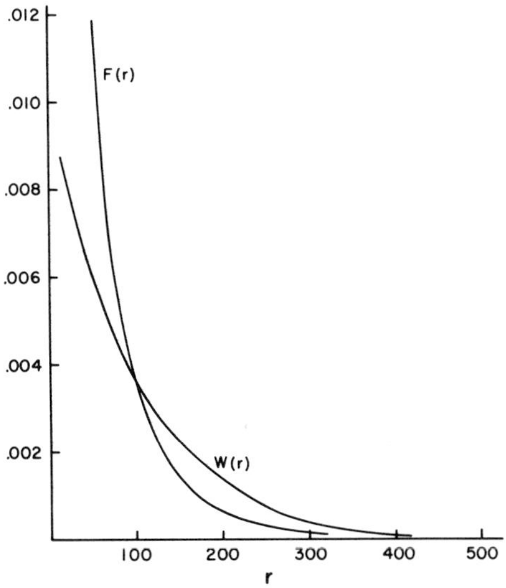

Figure 6.1.1 - Weight and frequency distribution for the equilibrium polymer polymerized to a randomly ordered state, $\bar{r}_w = 100$, $\bar{r}_n = 21.5$ (after Lundberg, 2).

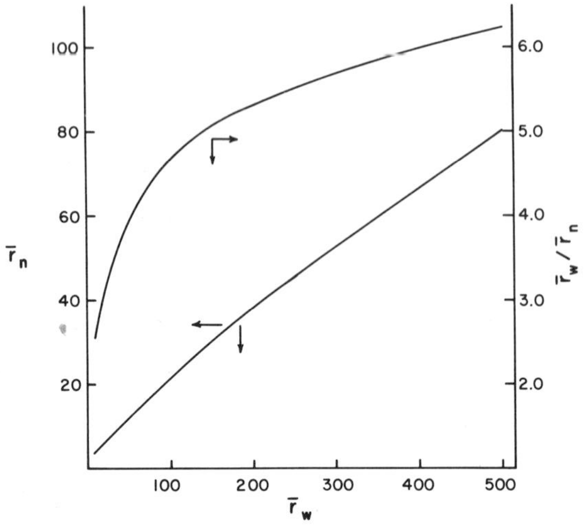

Figure 6.1.2 - $\bar{r}_n$ and $\bar{r}_w/\bar{r}_n$ as a function of $\bar{r}_w$ for the polymer of Figure 6.1.1. (After Lundberg, 2).

theory at finite concentrations. In fact Harris (3) points out that the Schulz-Flory distribution results when deviations from ideality are expressed in terms of the second virial coefficient, provided that the latter is independent of molecular weight. However, the distribution law is not independent of the thermodynamics of the system and indeed may vary significantly at high concentrations. Thus, as stated in Sections A and B, the problem revolves about the choice of the entropy associated with the chain configurations in the pure polymeric phase.

2. The Most Probable Distribution

In addition to the thermodynamic distribution already

discussed, the Schulz-Flory most probable distribution is also obtained for condensation polymerization when all the reactions are assumed to have the same probability regardless of whether or not exchange reactions occur:

$$P_r + P_s \rightleftharpoons P_{r+s-i} + P_i \qquad i < r + s \tag{1}$$

Here P_i can be a by-product, such as water, or a new polymer molecule whose size is smaller than r + s (Flory, 4, 5). The same distribution results when random scission occurs to infinitely long chains (Kuhn, 6).

3. <u>When Initiation, Propagation, and Termination are all Equilibrium Reactions</u>

Consider the system (Tobolsky, 7):

$$XY \rightleftharpoons X^+ + Y^- \qquad K_o, \text{ ionization of catalyst}$$

$$X^+ + M \rightleftharpoons MX^+ \qquad K_i, \text{ initiation}$$

$$XM_r^+ + M \rightleftharpoons XM_{r+1}^+ \qquad K_p, \text{ propagation}$$

$$XM_r^+ + Y^- \rightleftharpoons XM_rY \qquad K_t, \text{ termination}$$

The concentration of polymer molecules is

$$P_r = K_oK_iK_tXM(K_pM)^{r-1}$$

and the frequency function is

$$F(r) = (1 - K_p M)(K_p M)^{r-1}$$

which is the Schulz-Flory equation with $p = K_p M$. Furthermore,

$$\bar{r}_n = 1/(1 - K_p M) = (M_o - M)/(X_o - X)$$

where $\underline{X}$ is the equilibrium concentration of $\underline{XY}$. The same formulas are obtained from the following scheme:

$$XY + M \rightleftharpoons XMY \qquad K_a$$

$$XM_{r-1}Y + M \rightleftharpoons XM_rY \qquad K_p$$

where

$$K_a = K_o K_i K_t$$

Also

$$M_o = M(1 + K_a X \bar{r}_n^{\,2})$$

$$X_o = X(1 + K_a M \bar{r}_n)$$

When no initiator is necessary, we have the following two cases:

Case A

$$M \rightleftharpoons M^* \qquad K_a$$

$$M^*_{r-1} + M \rightleftharpoons M_r^* \qquad K_p$$

$$M_o = M(1 + K_a \bar{r}_n^{\,2})$$

$$\bar{r}_n = 1/(1 - K_p M)$$

$$\approx [(M_o K_p - 1)/K_a]^{\frac{1}{2}}$$

Case B

$$M + M \rightleftharpoons M_2 \qquad K_a$$

$$M_{r-1} + M \rightleftharpoons M_r \qquad K_p$$

$$M_o = M(1 + K_a M \bar{r}_n^{\,2})$$

$$\bar{r}_n = 1/(1 - K_p M)$$

$$\approx [M_o K_p^{\,2} - K_p)/K_a]^{\frac{1}{2}} \qquad K_a \neq K_p$$

$$\approx (M_o/M)^{\frac{1}{2}} \qquad K_a = K_p$$

4. Constant Number of Moles

If the catalyst can form an active species with any polymer molecule, and the total number of moles of the system is kept constant by continuously adding monomer (constant pressure and volume if all polymer molecules are ideal gases), then

P_o = the total number of moles in the system

P_1 = the number of monomers present at time *t*

$$P_r + C \overset{K}{\rightleftharpoons} P_r C$$

$$P_r C + P_1 \rightarrow P_{r+1} C \qquad k_p$$

$$1/P_1 - 1/P_o = k_p KCt$$

$$P_r = [P_1/(r - 1)!][\ln(P_o/P_1)]$$

(Fontana, 8)

$$\bar{r}_n = 1 + \ln (P_o/P_1)$$

$$\bar{r}_w = \frac{1 + 3 \ln (P_o/P_1) + [\ln (P_o/P_1)]^2}{1 + \ln (P_o/P_1)}$$

5. Equilibrium Polymerization with a Multifunctional Initiator

$X_f + M \overset{K}{\rightleftharpoons} X_f(M_1)$ initiation at one site, f - 1 unused sites left

$X_f(M_1) + M \overset{K}{\rightleftharpoons} X_f(M_1)(M_1)$ initiation at another site, f - 2 unused sites left

$X_f(M_1)(M_1) + M \overset{K_p}{\rightleftharpoons} X_f(M_2)(M_1)$ propagation at one site

etc.

Let

$y = 1 - K_p M$ and $z = KM$

N_s = number of molecules having s arms or branches ($s \leqslant f$)

$$= Xz^s / \sum_{i=1}^{s} [1 - (1 - y)^i]$$

$$= Xz^s/s!y^s \qquad \text{if } y \ll 1$$

M_s = number of monomers in molecules having s arms

$$= X\left[\frac{z}{1 - y}\right]^s \left\{ \prod_{i=1}^{s} [1 - (1 - y)^i]^{-1} \left[\frac{s}{1 - (1 - y)^s} + \sum_{i=1}^{s-1} \frac{(s - i)(1 - y)^s}{[1 - (1 - y)^{s-i}]}\right]\right\}$$

$$= Xz^s/[(s - 1)!y^{s+1}] \qquad y \ll 1$$

(Baur and Eisenberg, 9)

$$\bar{P}_n = z/y^2 [1 - \exp(-z/y)]$$

provided that $f \gg z/y$, that is, a polymer is used as initiator.

6. Polyphosphate Equilibria

Consider the following structures in a polyphosphate and the lowercase letters which represent these structures:

$$2 \begin{matrix} O \\ -OPO- \\ O \\ M \end{matrix} \underset{}{\overset{K_1}{\rightleftharpoons}} \begin{matrix} O \\ -OPO- \\ O \\ ' \end{matrix} + \begin{matrix} O \\ -OPOM \\ O \\ M \end{matrix} \qquad K_1 = be/m^2$$

$$2 \underset{m}{\text{middles}} \rightleftharpoons \underset{b}{\text{branch}} + \underset{e}{\text{end}}$$

$$2\ \mathrm{MO}\overset{\mathrm{O}}{\underset{\underset{\mathrm{M}}{\mathrm{O}}}{\mathrm{P}}}\mathrm{O}\text{-} \overset{K_2}{\rightleftharpoons} \text{-}\mathrm{O}\overset{\mathrm{O}}{\underset{\underset{\mathrm{M}}{\mathrm{O}}}{\mathrm{P}}}\mathrm{O}\text{-} + \mathrm{MO}\overset{\mathrm{O}}{\underset{\underset{\mathrm{M}}{\mathrm{O}}}{\mathrm{P}}}\mathrm{OM} \qquad K_2 = mo/e^2$$

$$\underset{e}{2\ \text{ends}} \rightleftharpoons \underset{m}{\text{middle}} + \underset{o}{\text{ortho}}$$

$$2\ \mathrm{MO}\overset{\mathrm{O}}{\underset{\mathrm{O}}{\mathrm{P}}}\mathrm{OM} \overset{K_3}{\rightleftharpoons} 2\ \mathrm{MO}\overset{\mathrm{O}}{\underset{\mathrm{O}}{\mathrm{P}}}\mathrm{O}\text{-} + [M_2O] \qquad K_3 = e^2u/o^2$$

$$\underset{o}{2\ \text{orthos}} \rightleftharpoons 2\ \underset{e}{\text{ends}} + \underset{u}{\text{unreacted}}\ M_2O$$

$$2\ \mathrm{MO}\overset{\mathrm{O}}{\underset{\underset{\mathrm{M}}{\mathrm{O}}}{\mathrm{P}}}\mathrm{OM} \overset{K'_3}{\rightleftharpoons} \mathrm{MO}\overset{\mathrm{O}}{\underset{\underset{\mathrm{M}}{\mathrm{O}}}{\mathrm{P}}}\mathrm{O}\overset{\mathrm{O}}{\underset{\underset{\mathrm{M}}{\mathrm{O}}}{\mathrm{P}}}\mathrm{OM} + [M_2O] \qquad K'_3 = 2eu/o^2$$

$$2\ \text{orthos} \rightleftharpoons \text{pyro} + \text{unreacted}\ M_2O$$

The $\underline{m}$, $\underline{b}$, $\underline{e}$, $\underline{o}$, and $\underline{u}$ are the mole fractions of these units present and

$$m + b + e + o + u = 1$$

Let $\underline{R}$ be the ratio of M_2O to P_2O_5; then

$$R = \frac{3o + 2e + m + 2u}{b + m + e + o}$$

for $R \geq 1$, $b = 0$ and

$$\bar{r}_n = (2m/e) + 2$$

$$W(r) = \frac{r}{\bar{r}_n(\bar{r}_n - 1)}\left[\frac{\bar{r}_n - 2}{\bar{r}_n - 1}\right]^{r-2}\left[1 - \frac{o}{1 - u}\right] \quad \bar{r}_n \geqslant 2,\ r \geqslant 2$$

(Van Wazer, 10).

7. Addition Polymerization Without Termination

A. Rate of Initiation equals $k_i MI$, no transfer, no termination, two active species in equilibrium, monomer concentration varies; see Section 3.4.

B. Simultaneous polymerization and depolymerization, no transfer, no termination; see Section 3.5.

8. References

1. K. F. O'Driscoll, "Equilibrium Polymerization," in Encyclopedia of Polymer Science and Technology, Interscience, New York, 1967, Vol. 6, p. 271.
2. J. L. Lundberg, "Molecular Weight Distribution in Equilibrium Polymerizations," J. Polymer Sci., A2, 1121 (1964).
3. F. E. Harris, "Equilibrium Distribution of Molecular Weights in Noncyclic Polymerizations," J. Polymer Sci., 18, 351 (1955).
4. P. J. Flory, Principles of Polymer Chemistry, Cornell University Press, Ithaca, N. Y., 1953.
5. P. J. Flory, "Fundamental Principles of Condensation Polymerization," Chem. Rev., 39, 137 (1946).

6. W. Kuhn, "Über die Kinetik des Abbau Hochmolekularer Ketten," Berichte, 63, 1503 (1930).
7. A. V. Tobolsky, "Equilibrium Polymerization in the Presence of an Ionic Initiator," J. Polymer Sci., 25, 220 (1957); 31, 126 (1957). A. V. Tobolsky and A. Eisenberg, "Equilibrium Polymerization of ε-Caprolactam," J. Am. Chem. Soc., 81, 2302 (1959); "General Treatment of Equilibrium Polymerization," 82, 289 (1960); "Transition Phenomena in Equilibrium Polymerization," J. Colloid Sci., 17, 49 (1962).
8. C. M. Fontana, in P. H. Plesch, The Chemistry of Cationic Polymerization, MacMillan, New York, 1963.
9. M. E. Baur and A. Eisenberg, "Equilibrium Polymerization. Polyfunctional Initiators," J. Chem. Phys., 42, 85 (1965).
10. J. R. Van Wazer, Phosphorus and Its Compounds, Vol. I, Interscience, New York, 1958.

Glossary of Symbols

a — The exponent in the equation $[\eta] = K\bar{r}_v^{\,a}$; otherwise as defined

$\bar{b}_n$, $\bar{b}_n(r)$ — Number-average number of branches per molecule for the entire distribution or for molecules of size _r_

c — Conversion

c_f — Conversion at infinite time, that is, polymerization ceases

C_m — Transfer constant to monomer, $k_{tr,m}/k_p$

C_p — Transfer constant to polymer, $k_{tr,p}/k_p$

C_s — Transfer constant to solvent, $k_{tr,s}/k_p$

Ei(+x) — The exponential integral, $\int_{-\infty}^{x} (e^u/u)\,du$

Ei(-x) — The exponential integral, $-\int_{x}^{\infty} (e^{-u}/u)\,du$

f — Efficiency of initiation; functionality of a system

$\bar{f}$ — Average functionality

F(r) — Frequency function or mole fraction of molecules of size _r_

G — Concentration of glycol in the polyester reaction. See Section 4.3

I — Initiator concentration

I_o — Initial initiator concentration

I* — Concentration of active initiator

$\tilde{I}_0$, $\tilde{I}_1$ — Modified Bessel function of the first kind of zero or first order, respectively

I(r) — $\int_0^r W(r)\,dr$, cumulative weight distribution

$\bar{j}_n$ Number-average number of coupling agents per molecule

k Exponent in the Schulz distribution [equation (1.3.2)] and in the generalized exponential distribution [equation (1.4.1)]

K As defined

k_d Specific rate constant for decomposition of initiator

k_i Specific rate constant for initiation

k_i' Specific rate constant for deactivation of active monomer [equation (3.5.1)]

k_{iex} Specific rate constant for the initiator expulsion reaction

k_{ij} Specific rate constant for radical type i adding to monomer type j

k_p Specific rate constant for propagation

k_p' Specific rate constant for depolymerization [equation (3.5.1)]

$k_{p,p}$ Specific rate constant for propagation through a polymer terminal double bond

$k_{s,i}$ Specific rate constant for initiation by a solvent radical

$k_{s,s}$ Specific rate constant for dimerization of two solvent radicals

k_t Specific rate constant for first-order termination or for second-order termination, where the mechanism is not specified

$k_{t,c}$ Specific rate constant for termination by combination, where two active centers form a single molecule

$k_{t,d}$ Specific rate constant for termination by disproportionation, where two active centers form two molecules by transfer

$k_{t,m}$ Specific rate constant for second-order termination with monomer

$k_{tr,m}$, $k_{tr,p}$, $k_{tr,s}$ Specific rate constants for transfer to monomer, polymer, or solvent, respectively

$k_{t,s}$ Specific rate constant for termination by a solvent radical

K_0, K_1, K_{11}, etc. Equilibrium constants for specified reactions

$\widetilde{K}_0$, $\widetilde{K}_1$ Modified Bessel function of the second kind of zero or first order, respectively

ℓ A running index; used to avoid confusion between the letter 1 and number 1

log p Common logarithm of p

ln p Natural logarithm of p

m A parameter in the generalized exponential distribution [equation (1.4.1)]

M, M_o Monomer concentration, initial monomer concentration

$M_1{}^*$ Concentration of active monomer formed by the act of transfer. It may contain a double bond capable of reacting at a later time

M_f Monomer concentration at infinite time

$\overline{M}_n$, $\overline{M}_w$ Number- or weight-average molecular weight

n Number of active initiator fragments obtained by decomposition of the initiator molecules

$\{n_s\}$ A set of numbers having a variety of subscripts $\underline{s}$ (Section 5.6 and 5.7)

N_o Initial concentration of monomer

p Extent of reaction

p_o, q_o Average composition of a copolymer chain, see Section 2.11

P_r Concentration of molecules of size $\underline{r}$

$P(x) = 1 - Q(x)$ Normal curve [equation (3.2.7)]

$P(\chi^2|\nu) = 1 - Q(\chi^2|\nu)$ Chi-square probability function, see Sections (3.2) and (3.7)

Q_r Concentration of molecules of size $\underline{r}$; the rth moment of P_n, $\Sigma n^r P_n$

r Number of monomer units in a polymer of size $\underline{r}$

$\binom{r}{i}$ $r!/(r - i)!i!$

$\bar{r}$ Mean of a distribution [equation (1.5.3)]

r_1, r_2 Reactivity ratios, in copolymerization

$\bar{r}_i$ Average of a distribution determined from the moments μ_i/μ_{i-1}

$\bar{r}_m$ Median value of a distribution [equation (1.6.1)]

$\bar{r}_n$ Number-average degree of polymerization [equation (1.1.14)]

$\bar{r}_n^o$ Initial number-average degree of polymerization

$\bar{r}_n|_b$, $\bar{r}_w|_b$ Number- and weight-average degrees of polymerization for molecules containing $\underline{b}$ branches

$\bar{r}_n(t = \infty)$, $\bar{r}_w(t = \infty)$ Values of $\bar{r}_n$ and $\bar{r}_w$ at infinite time

$\bar{r}_s$ A parameter in the generalized logarithmic normal distribution function [equation (1.6.4)]

$\bar{r}_v$ Viscosity-average degree of polymerization [equation (1.1.21)]

$\bar{r}_w$ Weight-average degree of polymerization [equation (1.1.16)]

$\bar{r}_z$ z-Average degree of polymerization

R Defined in the Gold distribution [equation (3.2.9)]

R^* Total concentration of active radicals = $\sum_{r=1}^{\infty} R_r^*$

R_o^* Total initial concentration of active radicals

R_I Rate of initiation

R_r, R_r^* Concentration of active radicals of size r

s A parameter in the generalized logarithmic normal distribution function [equation (1.6.4)]; otherwise, as defined

S Solvent concentration

S^* Concentration of active solvent formed by transfer

t Time; otherwise as defined

u, v Parameters in several distribution functions; they are defined each time they are used

w_1, w_2 Weight fraction of material of distribution 1, etc.

W(r) Weight fraction of molecules of size r

W(r, b) Weight fraction of molecules containing r units and b branches

$W_s(r)$ Generalized logarithmic normal distribution of parameter s [equation (1.6.4)]

W_{AA}, etc. Molecular weight of molecule AA, etc., it may be the molecular weight of the monomer or the repeat unit

x Relative molecular weight $= r/\bar{r}_n$; otherwise as defined

y A parameter in the generalized exponential distribution function [equation (1.4.1)]; otherwise as defined

y_o, y, y_1 Various functions of the relative initiator concentration (Sections 2.9 and 2.10)

Y_n $\sum_{r=1}^{\infty} r^n R_r^*$, the nth moment of R_r^*; otherwise as defined

z As defined

α, β, γ Extent of reaction of group type A, B, C,

α_c Critical conversion, the conversion at incipient gelation

β Fraction of monomers in branches (Section 5.4); otherwise, as defined

β', γ', ... Extent of reaction of a group which must be preceded by some other type of reaction, that is, a lactam ring must open before the amine can react

β_I, β_{II} Fraction of deactivated chains, see Section 3.3B

γ A parameter in several distribution functions; it is defined each time it is used

γ k_p/k_i for systems without termination; the cross-linking index

$\Gamma(a)$ Gamma function of a

δ, ϵ As defined

ζ A parameter in several distribution functions; it is defined each time it is used

η As defined

$[\eta]$ Intrinsic viscosity of a polymer

θ Dwell time in a continuous-flow reactor, equal to capacity divided by flow rate; as defined

$\varkappa$ As defined

λ, Λ A parameter in several distribution functions; it is defined each time it is used

Λ_0 Number of polymer molecules formed

Λ_1 Number of monomer molecules reacted, equal to the weight of polymer molecules formed

Λ_2 Second moment of the distribution

μ A parameter in several distribution functions; it is defined each time it is used

μ_n nth Moment of a distribution [equation (1.1.5)]

ν The expected value of the Poisson distribution [equation (1.7.1)]; the cyclomatic number of a system containing rings (Section 5.5); as defined

ξ A parameter in several distribution functions; it is defined each time it is used

$\prod_{i=0}^{n} A_i$ $A_0A_1A_2 \cdots A_{n-1}A_n$

ρ The branching density (Section 5.5); otherwise as defined

σ^2, σ Variance of a distribution [equation (1.1.9) also equations (1.5.1) and (1.6.1)]; otherwise as defined

τ A reduced time variable; as defined

φ As defined

$\varphi(\alpha', s)$ — $\sum_{n=1}^{\infty}(\alpha')^n n^{-s}$, see Table 5.14.1

Φ — A parameter in several distribution functions; it is defined each time it is used

ψ — As defined

$\Omega(r)$ — Fraction of molecules formed by transfer of size r [equation (2.2.4)]

Author Index

Citation in brackets gives the page and reference number, the latter enclosed in parentheses.

Subject Index

References to specific kinetic cases are made in the contents section of each chapter.